Steffen Heldt

Near-optimal operation of LNG liquefaction processes

Steffen Heldt

Near-optimal operation of LNG liquefaction processes

by means of regulation

Südwestdeutscher Verlag für Hochschulschriften

List of Tables

$\boldsymbol{\nu}_i$ Eigenvector corresponding to the ith eigenvalue

η Dynamic viscosity (in $\mathrm{kg\,m^{-1}\,s^{-1}}$)

π Compression pressure ratio (> 1)

ρ Density (in $\mathrm{kg\,m^{-3}}$)

$\boldsymbol{\Sigma}$ Diagonal matrix

σ Frictional tension (in $\mathrm{kg\,m^{-1}\,s^{-2}}$)

σ_i i^{th} smallest/largest (depending on context) singular value

τ Integral time parameter of PID controller

θ Correction term for mass transfer

ω Angular frequency (in $\mathrm{rad\,s^{-1}}$)

ζ Dimensionless longitudinal coordinate

Latin symbols

A Area (in $\mathrm{m^2}$)

a Thermal diffusivity (in $\mathrm{m^2\,s^{-1}}$)

$\mathbf{c}$ CV vector

$\boldsymbol{c}$ Molar density vector (in $\mathrm{mol\,m^{-3}}$)

C_{f} Enhancement multiplier for the effect of liquid film turbulence

C_k^n Binomial coefficient indicating number of possibilities for selecting k among n elements

c_p Specific heat capacity (in $\mathrm{m^2\,s^{-2}\,K^{-1}}$)

$\mathbf{d}$ DV vector

$\mathcal{D}$ DV space

D Coil diameter (in m)

d Diameter (in m)

d_{c} Outer tube diameter (in m)

d_{i} Inner tube diameter (in m)

E Worst-case expectation value

e Mass specific internal energy (in $\mathrm{m^2\,s^{-2}}$)

h Mass specific enthalpy (in $\mathrm{m^2\,s^{-2}}$)

$\boldsymbol{F}$ Matrix indicating $\partial \boldsymbol{y}_{\mathrm{opt}}/\partial \boldsymbol{d}$

$\boldsymbol{F}$ Molar flow rate vector (in $\mathrm{mol\,s^{-1}}$)

F Molar flow rate (in $\mathrm{mol\,s^{-1}}$)

f Empirical correlation (dimension from context)

f Frequency (in $\mathrm{s^{-1}}$)

F_V Volumetric flow rate (in $\mathrm{m^3\,s^{-1}}$)

G Mass flux (in $\mathrm{kg\,s^{-1}\,m^{-2}}$)

g EOS for mass specific enhalpy (in $\mathrm{m^2\,s^{-2}}$); subscripts p and $\boldsymbol{c}$ refer to partial

	derivatives
g	Gravitational acceleration $\left(\text{in m s}^{-2}\right)$
$\boldsymbol{G}_{\boldsymbol{w}}^{\boldsymbol{v}}$	Transfer matrix from $\boldsymbol{w}$ towards $\boldsymbol{v}$ (for the sake of simple notation, the indication of $\boldsymbol{v} = \boldsymbol{c}$ or $\boldsymbol{w} = \boldsymbol{u}$ is omitted)
$\boldsymbol{\mathfrak{H}}$, $\boldsymbol{H}$	Non-linear and linear map, representing $\partial \boldsymbol{c}/\partial \boldsymbol{y}$
H	Enthalpy $\left(\text{in kg m}^2\,\text{s}^{-2}\right)$
h	Mass specific enhalpy $\left(\text{in m}^2\,\text{s}^{-2}\right)$
$\boldsymbol{h}_i$	i^{th} row of $\boldsymbol{H}$
J	Cost function
j	Flux $\left(\text{in m s}^{-1}\right)$
j	Imaginary number $\sqrt{-1}$
$\boldsymbol{L}$	Loop transfer function
L	Loss
L	Length (in m)
$\boldsymbol{M}$	Loss matrix
$\tilde{\boldsymbol{M}}$	Molar mass vector $\left(\text{in kg mol}^{-1}\right)$
M	Mass (in kg)
$\mathcal{N}(\mu_1, \mu_2)$	Normal distribution with mean μ_1 and variance μ_2
n_{c}	Number of components
$\boldsymbol{n}^{\boldsymbol{c}}$, $\boldsymbol{n}^{\boldsymbol{y}}$	Implementation errors relating to input and output uncertainty
n_{r}	Speed of rotation (in RPM)
$n_{\boldsymbol{v}}$	Size of vector $\boldsymbol{v}$
$n_{\mathcal{Y}}$	Size of PV subset $\mathcal{Y}$
$\mathcal{O}$	Landau notation (describes limiting behavior of a function)
$\boldsymbol{P}$	Rosenbrock matrix
p	Pressure $\left(\text{in kg m}^{-1}\,\text{s}^{-2}\right)$
P_{l}, P_{r}	Longitudinal, radial distance of tubes (in m)
q	Volume specific heat $\left(\text{in kg m}^{-1}\,\text{s}^{-2}\right)$
Q	Heat duty $\left(\text{in kg m}^2\,\text{s}^{-3}\right)$
$\boldsymbol{S}$	Sensitivity function
S	Slip ratio
$\boldsymbol{T}$	Complementary sensitivity function
T	Temperature (in K)
t	Time (in s)
$\boldsymbol{U}$	Unitary matrix
$\boldsymbol{u}$	MV vector
$\mathcal{U}(x_1, x_2)$	Continuous uniform distribution in the interval $[x_1, x_2]$

Disclaimer

MATHEMATICA® is a registered trademark of Wolfram Research, Inc., Champaign, Illinois. MATLAB® is a registered trademark of The Mathworks, Natick, Massachusetts. NAG® is a registered trademark of The Numerical Algorithms Group, Oxford, United Kingdom. OPTISIM®, MFC® and LIMUM® are registered trademarks of the Linde AG, Munich, Germany. Intel® Core™ is a trademark of the Intel Corporation, Santa Clara, California. WINDOWS XP® is a registered trademark of the Microsoft Corporation, Redmond, Washington. AP-X™ is a trademark of Air Products and Chemicals, Inc., Allentown, Pennsylvania.

1. Introduction

This chapter gives an outline of the thesis. In Section 1.1, technical solutions for optimal operation of process plants are presented. A particular solution is selected and it is reasoned why it is considered favorable. Background information regarding liquefied natural gas (LNG) is supplied in Section 1.2. Previous work strongly related to this thesis is introduced in Section 1.3. Section 1.4 summarizes the main contributions of this work. The structure of the thesis is outlined in Section 1.5.

1.1. Motivation

Process plants are designed for a certain set of process conditions such as feed and ambient conditions where they must achieve targets as required by the product specifications (*e.g.*, product purity). During operational practice, the plant must be manipulated in order to satisfy the targets irrespective of varying conditions. Therefore, each target is related to a controlled variable (CV) which is kept at its setpoint variable (SV) by a controller which therefore acts on one or more manipulated variables (MVs) such as a valve position, speed of rotational equipment, *etc.* If there are less MVs than targets, then the targets cannot be achieved simultaneously due to too few available degrees of freedom. In case of as many MVs as targets, the targets may be achieved with one possible adjustment of the MVs which can be found by the controller(s). If there are more MVs than targets, then the targets may be achieved by various adjustments of the MVs. Practically spoken, the MVs can be separated into one set which is mapped via controller loops towards the target related CVs and another set called surplus MVs whose values can be independently set. That the surplus MVs are seldom independent, *i.e.*, unused by controllers, relates to the fact that most engineers use the surplus MVs to keep other process variables (PVs) fixed for the sake of smoother operation, higher reliability, *etc.* It is important to stress that the additional degrees of freedom are generally not lost by using the surplus MVs but simply moved from the surplus MVs to the SVs of the additionally introduced controller loops which are referred to as surplus SVs.

The significance of the above stated is that a plant with surplus MVs/SVs can be subject to steady-state optimization of an economic measure such as profit and cost. If the steady-state behavior of the plant is strictly linear, then a linear programming problem

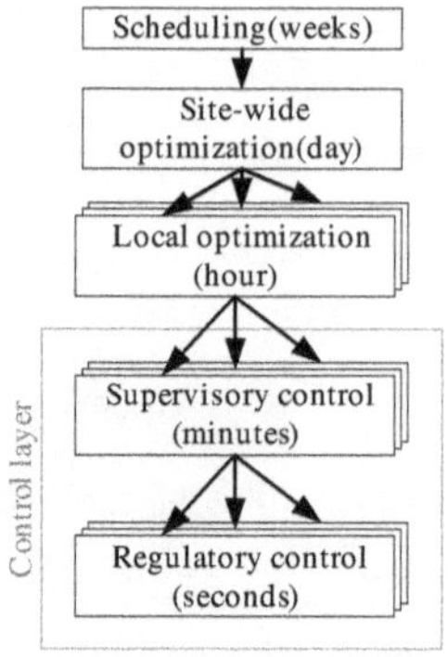

Figure 1.1.: Typical control hierarchy of a chemical plant (after Skogestad, 2000)

must be solved and the optimum is located directly at the edge of the operating region. Usually linear model predictive control (LMPC) is utilized in practice to achieve the solution to the linear programming problem. Note that LMPC takes also plant dynamics into account and provides smooth transitions from one steady-state operating point to another. If the assumption of a linear steady-state behavior is inadequate, then a nonlinear programming problem needs to be solved and the optimum can be anywhere within the operating region. Real-time optimization (RTO) is the most common technology to obtain the optimal steady-state operating point of nonlinear plants.

Advanced control technologies such as LMPC or RTO are not considered in this work for the investigation of optimal operation of LNG liquefaction plants. This is due to various reasons but mostly related to the fact that this work is designated to serve the needs of a plant contractor such as the Engineering Division of the Linde AG. The Linde Engineering Division is particularly interested in solutions which can be installed in an early phase of the life cycle of a plant, which eases the commissioning of plants and helps to achieve agreed specifications during the performance tests, which can be reused among similar projects and which may be sold as a special accessory to the plant. These requirements are generally not achieved by advanced control technologies such as LMPC and RTO. For instance, the model for LMPC is obtained by step tests of the plant which is time-consuming and cannot be performed until commissioning has taken place. Further, the model is individually related to a particular project, even for a certain phase of the life cycle, and cannot be reused among projects. In contrast, RTO is based on a rigorous first-principle models of the plant which may by adapted between projects if justified by a cost-benefit analysis. However, taking RTO into operation relies also on a already commissioned plant. Consequently, its rather time-consuming implementation comes earliest into play after commissioning.

Remark 1.1. Some plant operators, especially in market segments such as ethylene plants where advanced process control is traditionally applied (Seborg, 1999; Friedman, 1999; Kano and Ogawa, 2009), desire the implementation of such technologies as early as possible. Therefore efforts have been recently undertaken by the Linde Engineering Division to integrate the development of advanced control projects into the earliest phase of the life cycle of a plant to make the benefits for the customer available as soon as possible (Schulze, 2007/11/18-20).

Self-optimizing control (Skogestad, 2000) is considered the approach which serves the needs of a plant contractor best. It has nothing to do with online optimization as its name may suggest. It refers to simple regulatory control where only the set of surplus CVs is specially selected such that the plant is operated near its optimal state despite varying disturbances and fixed SVs. Regulatory control is utilized in the lower layer of virtually every state-of-the-art plant as indicated in Figure 1.1. Self-optimizing control is therefore easy to implement and practically accepted by plant operating staff. The arrows in the sketch of Figure 1.1 indicate that overlying layers determine the SVs of underlying ones. Advanced technologies such as LMPC and RTO are classified in the local and site-wide optimization layer, respectively. It is thus clear that self-optimizing control does not exclude advanced technologies but can be commonly applied with them. It is interesting to note that as a part of his thorough survey of optimal operation by feedback control, Engell (2006/04/02-05) proposes the re-thought of the traditional layer functionality as indicated in Figure 1.1. He suggests taking better advantage of regulatory and advanced control concepts by a higher integration of both worlds.

The fact that economically optimal operation of a plant can be achieved by regulatory control and a constant setpoint policy is by far not obvious and is therefore illustrated by an example.

Example 1.2. Suppose that an exemplary plant with one surplus MV and two candidate CVs $\boldsymbol{y} = \begin{bmatrix} y_1 & y_2 \end{bmatrix}^T$ is present where $\boldsymbol{y} = \boldsymbol{0}$ holds at the nominal point. Suppose further that the cost function of the plant to be minimized during operation is given by

$$J = \boldsymbol{y}^T \begin{bmatrix} 5 & -1 \\ -1 & 0.1 \end{bmatrix} \boldsymbol{y}.$$

By observation of the contour plot of the cost function presented in Figure 1.2, it is fairly apparent that it is economically more favorable to keep y_1 at its nominal value than y_2. An even better CV selection is the linear combination $y_2 - 4.2361\, y_1$ indicated by the dashed line.

The example shows vividly that the appropriate selection or combination of CVs can

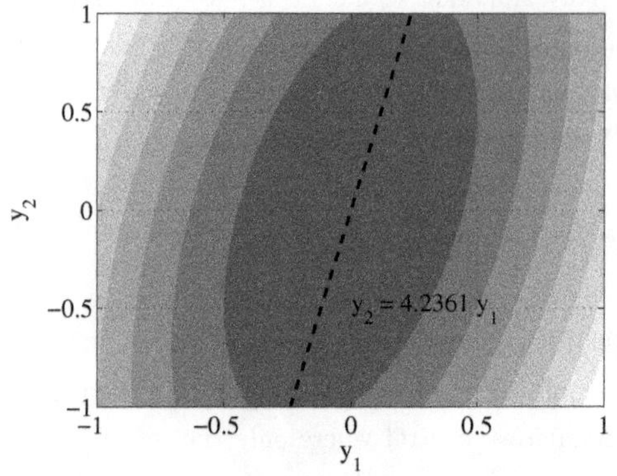

Figure 1.2.: Cost of an exemplary plant

have an effect on plant economics if the SVs are intended to be fixed. What was not pointed out is how CVs are selected in order to obtain self-optimizing control behavior. Generally, control engineering of the two lowermost layers in Figure 1.1 involves five steps (Skogestad, 2004a):

1. Selection of MVs
2. Selection of CVs to be kept at certain SVs
3. Selection of CVs for stabilization of the plant
4. Selection of control configuration
5. Selection of controller type and design parameters

Take into account that the first three steps may be interchangeable. Mostly in industrial practice, all steps are performed based on experience and engineering insight, still very much along the lines described by Buckley (Apr 1965, Chapter 13). The second step is referred to as control structure design (CSD) and a systematic approach based on simulation models can be beneficial in terms of economic measures.

The fact that model-based control engineering is still rarely applied in industry is due to various reasons. First, in most cases there are simply no appropriate plant models. Second, control is usually a multi-objective task and minimum energy consumption aimed by self-optimizing CSD is usually not the most critical objective; the engineers often rather focus on operational simplicity. Third, the framework of self-optimizing CSD is fairly new and is just about to arrive at industrial practice.

It is worth thinking of how the steps above affect plant operation. The first step is responsible for plant versatility which refers to the ease of changing the operating point.

For instance, it generally holds that the more MVs a plant has, the more versatile is it. A simple example of an MV which can be added or removed is a bypass around a heat exchanger. The second step affects also the versatility and as pointed out above the optimality of the plant. The former can also be illustrated by an example.

Example 1.3. Suppose one selects a CV which is almost independent on the MVs. Then the versatility of the plant is poor as the MVs have limited range and do not manage to achieve the given SV due to saturation.

Most of the times, the third in the above listed steps is also very obvious. For instance the level in a separation drum must be stabilized by using either the inlet or one of the outlet flow rates as MVs. It is interesting to stress that thereby the degrees of freedom remain invariant as the lost in MVs is replaced by the degrees of freedom generated by the SVs. However, in the particular case of level control the SV has usually no steady-state effect. Control configuration in step four refers to the structure of the overall controller that interconnects the CVs, MVs and extra PVs. Thereby, dynamical properties such as stability, integrity, response promptness, oscillations, MV saturation *etc.* are affected. The fourth step is somewhat interconnected with the fifth step and vice versa. The control law specification such as PID, decoupler, *etc.* affect exclusively dynamical properties.

This thesis deals with the systematic control structure design (step 1 through 3), selection of control configuration (step 4) and controller synthesis (step 5) for selected LNG liquefaction processes. Here, the emphasis lies on the selection of CVs (step 2) with special regard to self-optimizing control.

1.2. Background of LNG

Liquefied natural gas (LNG) is a fully condensed methane-rich ($>$ 90 mol%) mixture of hydrocarbons and nitrogen near atmospheric pressure. LNG is present at a temperature of approximately $-160\,^{\circ}$C. The actual value depends on product pressure and composition. LNG is generated from natural gas free from traces of water and acid gas. The reasons for production of LNG is that its density is increased by a factor of 600 compared to natural gas at same pressure conditions yielding about 40 % more heating value than any liquid fuel derived from the chemical conversion of natural gas (Zaïm, Mar 2002, p. 1).

Due to this increase in density, transportation via fleets of tank ships and trucks becomes feasible. In some cases and aspects it can be advantageous over pipeline transport, in others it may be the only possibility. Figure 1.3 shows the transportation cost versus distance for ship and onshore/offshore pipeline transport. LNG shipping is the only transportation method which has nonzero cost at zero distance which is due to distance-independent investment cost. As the transportation cost per unit and distance are the

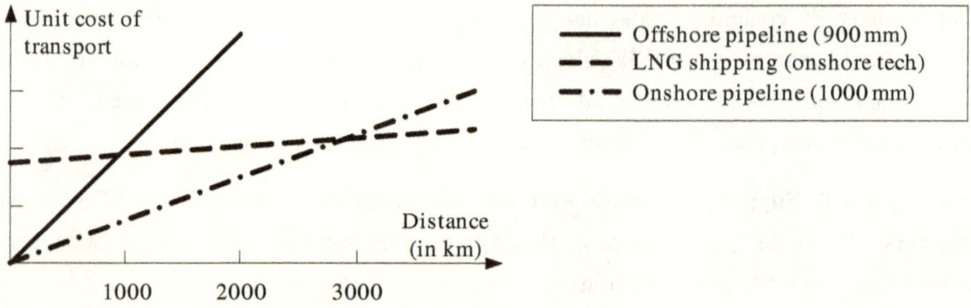

Figure 1.3.: Comparison of transportation cost (after Hammer *et al.*, 2006)

smallest for LNG shipping, it becomes economically favorable over both offshore and onshore pipeline transport at certain distances. It is therefore particularly useful for the exploration of remote gas fields. Another argument pro LNG transportation is that instead of delivering the total LNG to one customer, various customers which may be remote from the plant and remote from each other can be delivered. For example, LNG produced in the Middle East and throughout the Pacific Rim supplies approximately 10 % of Japan's primary energy consumption (Hammer *et al.*, 2006). Further, LNG can be traded as spot delivery which provides more flexibility as pipeline distribution for both the customer and the supplier. The production of LNG can also be useful for peakshaving purposes. For instance, storage of LNG near urban areas permits peak demands for natural gas to be satisfied without building additional pipelines that would be underutilized most of the time.

The first commercial shipping of LNG was realized in 1964 with the export from the La Camel LNG plant in Arzrew, Algeria to Canvey Islands, UK. Since then, the LNG technology has evolved in terms of higher efficiency and larger plant scales which both improved LNG project economics significantly (Yates, 2002/10/13-16; Berger *et al.*, 2003a). Recent activities of the Engineering Division of the Linde AG and other players aim to develop offshore LNG production plants, so called floating production storage and off-loadings (FPSOs), for the exploration of stranded gas reserves (Voskresenskaya, 2009/03/31-04/02).

LNG global trade has been expanding almost steadily since its beginning and the Asia-Pacific region is dominating the demand since the mid-seventies (Jensen, 2004). The current annual global demand is about 226 billion standard cubic meters (BCM) (BP stat. rev., Jun 2009). According to the forecast of McKay (Jan 2009), the LNG volume will increase to 340 BCM in 2015 and to 680 BCM in 2030 despite the current financial melt-down.

1.3. Previous work

Previous and related work in fields addressed by this thesis are specifically referred to in the respective chapters, *i.e.*, in Sections 2.1, 3.1, 4.1 and 5.1. Nevertheless, the difference to the closely related work of Jensen (May 2008) is particularly pointed out here. Jensen (May 2008) was the first which applied self-optimizing control structure design to simple refrigeration and heat pump cycles and two LNG liquefaction processes, the PRICO® cycle (Jensen and Skogestad, 2006/04/02-05, 2009b) and the MFC® process (Jensen and Skogestad, 2006/07/09-13). His aim was the identification of sets of controlled variables which automatically lead to almost optimal operation.

In this work, similar studies are performed for the MFC® process and the LIMUM® cycle with some major differences. First, a different objective function for optimal operation is used. In Appendix 4.A, it is pointed out that the LNG throughput is considered the most general objective in terms of independence from market conditions. Second, new control structure design methods called AM methods are proposed in this thesis. They allow an engineer to find the best set of linear combinations of process variables for each controlled variable independently. As these methods were not available for Jensen (May 2008), he restricted himself mostly to the search of the best selection of PVs rather than their combination. It is important to stress that PV combination structures generally achieve better optimality as PV selection structures. Third, in contrast to the work of Jensen (May 2008), the newly identified control structures are judged by means of dynamic operability metrics.

1.4. Contributions

Spiral wound heat exchangers (SWHEs) are commonly applied as main cryogenic heat exchangers in baseload LNG liquefaction plants. In order to provide accurate predictions of LNG plant behavior, an existing basic model of an SWHE was further developed into a highly sophisticated model and implemented into the Linde in-house simulator OPTISIM®. In contrast to models of prior art, the new model allows for temporal mass accumulation within the tube and shell passages. Further enhancements refer to the implementation of empirical correlations for tube and shell-side heat transfer and pressure drop in SWHEs with special regard to LNG service. This was rather challenging task as different stream regimes take place on both, the tube and the shell side.

Based on a linearized steady-state plant model and the second derivative (Hessian) of the cost/profit function, both obtained at the plant's nominal operating point, sets of controlled variables can be judged in terms of closeness to optimal plant behavior by the worst-case/average loss criterion (Halvorsen *et al.*, 2003; Kariwala *et al.*, 2008). The problem of finding the best set of controlled variables has been considered by various authors. Some considered the efficient selection of process variables as controlled variables (Cao and Saha, 2005; Kariwala and Skogestad, 2006/07/09-13; Cao and Kariwala, 2008; Kariwala and Cao, 2009, 2010a). Others derived methods in which every controlled variable is a linear combination of the same predefined set of process variables (Alstad and Skogestad, 2007; Kariwala, 2007; Kariwala *et al.*, 2008; Alstad *et al.*, 2009). Further enhancement of some of these methods allowed the efficient identification of the best process variable subset of a certain size (Kariwala and Cao, 2009, 2010a). An alternate slightly less efficient method accomplishes to take structural constraints for the process variable subset into account (Yelchuru and Skogestad, 2010/07/25-28).

From a practical point of view, structures where every controlled variable consist of the same process variable set have little practical acceptance. Furthermore, their optimality becomes insufficient if the process variable subset size is small (in the order of the controlled variable set size). Due to these deficiencies, efforts have been undertaken in this work to develop identification methods for more advanced combination structures. The results are methods that are able to identify structures in which each controlled variable can be a linear combination of an individual PV subset of predefined subset size. These methods have been already made publicly available (Heldt, 2009, 2010a) and are presented in this thesis.

The new control structure design methods were applied to three LNG processes, the simple SMR cycle, the LIMUM® cycle and the MFC® process. The most promising results were successfully proven by dynamic operability analyses and protected by patents (Heldt, pending).

1.5. Thesis structure

The structure of the thesis is outlined in the following. Chapter 2 deals with the modeling of LNG plants. The most importance lies on the model of the spiral wound heat exchangers as they are the major equipment of most baseload LNG plants. In Chapter 3, the concept of self-optimizing control is introduced and the problem of finding best controlled variable sets is discussed. Publicly available solution methods are reviewed and new advanced methods are proposed. In order to illustrate the advantage of the latter, they are compared with the former by repeated random tests. Their application to a academic process example and three LNG liquefaction processes takes place in Chapter 4. The results are new sets of controlled variables which provide almost optimal operation of the considered processes when a constant setpoint policy is applied. For the most promising controlled variable sets of the LIMUM® cycle and the MFC® processes, the best pairings with the manipulated variables are figured out in Chapter 5 by means of dynamical measures of the linearized dynamic model equations. The dynamical performance of these new control strategies is compared with conventional ones. Concluding remarks and outlook for future work are given chapter-specific in Sections 2.4.8, 3.6, 4.8, 5.6.

Bibliography

BP statistical review of world energy, Jun 2009.

V. Alstad and S. Skogestad. The null space method for selecting optimal measurement combinations as controlled variables. *Industrial and engineering chemistry research*, 46 (3):846–853, 2007.

V. Alstad, S. Skogestad, and E. S. Hori. Optimal measurement combinations as controlled variables. *Journal of process control*, 19(1):138–148, 2009.

E. Berger, W. Förg, R. S. Heiersted, and P. Paurola. Das Snøhvit-Projekt: Der Mixed Fluid Cascade (MFC(R)) Prozess für die erste europäische LNG-Baseload-Anlage. *Linde technology*, (1):12–23, 2003a.

P. S. Buckley. *Techniques of process control.* Krieger Publishing, Melbourne, Florida, Apr 1965. ISBN 0471116556.

Y. Cao and V. Kariwala. Bidirectional branch and bound for controlled variable selection: Part I. Principles and minimum singular value criterion. *Computers and chemical engineering*, 32(10):2306–2319, 2008.

Y. Cao and P. Saha. Improved branch and bound method for control structure screening. *Chemical engineering science*, 60(6):1555–1564, 2005.

S. Engell. Feedback control for optimal process operation. International symposium on advanced control of chemical processes, Gramado, Brazil, 2006/04/02-05.

Y. Z. Friedman. Advanced control of ethylene plants: What works, what doesn't and why. *Hydrocarbon Asia*, 9(Jul/Aug):1–14, 1999.

I. J. Halvorsen, S. Skogestad, J. C. Marud, and V. Alstad. Optimal selection of controlled variables. *Industrial and engineering chemistry research*, 42:3273–3284, 2003.

G. Hammer, T. Lübcke, R. Kettner, M. R. Pillarella, H. Recknagel, A. Commichau, H. J. Neumann, and B. Paczynska-Lahme. Natural Gas. In *Ullmann's encyclopedia of industrial chemistry: Electronic Release 2006.* Wiley-VCH, 2006. ISBN 3527313184.

S. Heldt. Verfahren zum Betrieb einer Anlage zum Verflüssigen eines kohlenwasserstoffreichen Stroms (Patent pending).

S. Heldt. On a new approach for self-optimizing control structure design. In S. Engell and Y. Arkun, editors, *ADCHEM 2009: Preprints of IFAC symposium on advanced control of chemical processes: July 12-15, 2009, Koç University, Istanbul, Turkey*, volume 2, pages 807–812. 2009.

S. Heldt. Dealing with structural constraints in self-optimizing control engineering. *Journal of process control*, 20(9):1049–1058, 2010a.

J. B. Jensen. *Optimal operation of refrigeration cycles: Ph.D. thesis.* Ph.D. thesis, NTNU, Trondheim, Norway, May 2008.

J. B. Jensen and S. Skogestad. Optimal operation of a simple LNG process. International symposium on advanced control of chemical processes, Gramado, Brazil, 2006/04/02-05.

J. B. Jensen and S. Skogestad. Optimal operation of a mixed fluid cascade LNG plant. Symposium on Process Systems Engineering/European Symposium on Computer Aided Process Engineering, Garmisch-Partenkirchen, Germany, 2006/07/09-13.

J. B. Jensen and S. Skogestad. Single-cycle mixed-fluid LNG process: Part I: Optimal design. In H. E. Alfadala, G. V. R. Reklaitis, and M. M. El-Halwagi, editors, *Proceedings of the 1st annual gas processing symposium: 10 - 12 January 2009, Doha, Qatar*, pages 213–220. Elsevier, 2009b. ISBN 9780444532923.

J. T. Jensen. *The development of a global LNG market: Is it likely? if so, when?* Oxford institute for energy studies, Oxford, 2004. ISBN 1901795330.

M. Kano and M. Ogawa. The state of the art in advanced chemical process control in Japan. In S. Engell and Y. Arkun, editors, *ADCHEM 2009: Preprints of IFAC symposium on advanced control of chemical processes: July 12-15, 2009, Koç University, Istanbul, Turkey*, volume 1, pages 11–26. 2009.

V. Kariwala. Optimal measurement combination for local self-optimizing control. *Industrial and engineering chemistry research*, 46(46):3629–3634, 2007.

V. Kariwala and Y. Cao. Bidirectional branch and bound for controlled variable selection: Part II. Exact local method for self-optimizing control. *Computers and chemical engineering*, 2009.

V. Kariwala and Y. Cao. Bidirectional branch and bound for controlled variable selection: Part III. Local average loss minimization. *IEEE transactions on industrial informatics*, 2010a.

V. Kariwala and S. Skogestad. Branch and bound methods for control structure design. Symposium on Process Systems Engineering/European Symposium on Computer Aided Process Engineering, Garmisch-Partenkirchen, Germany, 2006/07/09-13.

V. Kariwala, Y. Cao, and S. Janardhanan. Local self-optimizing control with average loss minimization. *Industrial and engineering chemistry research*, 47(4):1150–1158, 2008.

J. McKay. LNG output to surge and new projects face funding hurdles. *LNG journal*, pages 1–4, Jan 2009.

K. Schulze. Modellbasierte Prozessführung: Erfahrungen und Herausforderungen aus der Sicht eines Anlagenbauers. Prozess-, Apparate- und Anlagentechnik, Weimar, Germany, 2007/11/18-20.

D. E. Seborg. A perspective on advanced strategies for process control. *Automatisierungstechnische Praxis*, 41(11):13–31, 1999.

S. Skogestad. Plantwide control: The search for the self-optimizing control structure. *Journal of process control*, 10(5):487–507, 2000.

S. Skogestad. Control structure design for complete chemical plants. *Computers and chemical engineering*, 28(1-2):219–234, 2004a.

E. Voskresenskaya. Potential application of floating LNG. Global conference on renewables and energy efficiency for desert regions, Amman, Jordan, 2009/03/31-04/02.

D. Yates. Thermal efficiency: Design, lifecycle, and environmental considerations in LNG plant design. Gastech, Doha, Qatar, 2002/10/13-16.

R. Yelchuru and S. Skogestad. MIQP formulation for optimal controlled variable selection in self-optimizing control. The international symposium on design, operation and control of chemical processes, Singapore, Republic of, 2010/07/25-28.

A. Zaïm. *Dynamic optimization of an LNG plant: Case study GL2Z LNG plant in Arzew, Algeria: Ph.D. thesis*, volume 10 of *Schriftenreihe zur Aufbereitung und Veredelung*. Shaker, Aachen, Germany, Mar 2002. ISBN 3832206647.

2. Modeling

The investigation of the operation of LNG liquefaction processes is based on dynamical simulation models of these processes. The models used throughout this thesis are built in the Linde in-house simulator OPTISIM®. This Chapter gives an overview on property calculation for mixtures of hydrocarbons and governing equations for standard unit operations present in LNG liquefaction processes. Section 2.1 refers to prior work in the field of modeling of LNG liquefaction processes. Section 2.2 deals with the calculation of properties of natural gas and mixed refrigerants. In Section 2.3, a survey of modeling of unit operations present in LNG liquefaction processes is given. The further development of a spiral wound heat exchanger model is presented in Section 2.4. Some issues concerning the modeling of cycle processes are discussed in Section 2.5.

2.1. Related work

Dynamic modeling of LNG liquefaction processes has been discussed in various prior works, *e.g.*, Melaaen (Oct 1994); Zaïm (Mar 2002) and Hammer, all of them Ph.D. theses. Kronseder (2003) considered the dynamic modeling of chemical engineering plants in OPTISIM® for the sake of dynamic optimization, parameter estimation and online optimal control. A thorough introduction into process modeling is given by Najim (1989) and Hangos and Cameron (1997, 2001). Related work regarding the modeling of spiral wound heat exchangers is referred to in Sections 2.4.2 and 2.4.3.

2.2. Material properties

For physical property calculation, Linde has made significant investments into its in-house General Multiphase Property System (GMPS) package which is integrated into OPTISIM® among other tools (Burr and Pfeiffer, 1983; Burr, 1985; Steinbauer and Hecht, 1996/06/03-05). Similar to process simulation, there is a more than 20-year-old history of developing own physical property methods. With regard to technology, specific multiphase calculation capabilities are still outstanding compared to commercial offerings. Further advantages are given by the flexibility to rapidly react to new requirements, and by the internal expert know-how which allows highly reliable guarantees, *e.g.*, for product purities (Engl and Kröner, 2006/07/09-13).

For calculation of properties and phase equilibrium of hydrocarbon mixtures of components C1 to C6 and nitrogen, a proprietary form of the Soave-Redlich-Kwong (SRK) equation of state (EOS) was used in this work. The SRK EOS is a cubic EOS which is known to accurately describe non-ionic mixtures. This was verified by a recent study by Jerinic *et al.* (2009) which applied popular EOSs to natural gas mixtures and compared them with respect to their prediction accuracy of properties such as vapor pressure, dew/boiling density, compressibility factor as well as binary gas/liquid equilibrium and condensate fraction. They came to the conclusion that the similar Peng-Robinson EOS is best-suited for the prediction of the behavior of natural gas mixtures. It is worth mentioning that in other publicly available works where LNG liquefaction processes were modeled and simulated (*i.e.*, Melaaen, Oct 1994; Zaïm, Mar 2002; Hammer, pp. 35-37), exclusively the cubic EOSs of Peng-Robinson and SRK were applied.

A thorough description of physical behavior of natural gas systems is given in the textbook of Katz and Lee (1990, pp. 110-181). One interesting property of hydrocarbon mixtures is the (counterintuitive) retrograde condensation which takes place in the two phase region near the critical point as indicated by the gray area in the phase diagram in Figure 2.1.

2.3. Unit operations

Modeling of chemical plant processes takes place in a modular fashion and is commonly referred to as flowsheeting. Flowsheeting relates to incrementally building a graphical/code representation of a process by adding unit operations from templates and interconnecting those using material, energy and information streams. In the background, a system of equations is generated which represents the plant model. It may be a pure algebraic system for design or steady-state simulation purposes or, in case of a dynamic simulation model, a mix of differential-algebraic equations (DAEs). A survey of the technology of equation-oriented flowsheeting is given by Barton (Mar 2000). A more general overview of flowsheeting tools is given by Zerry (2008). The Linde in-house development OPTISIM®[1] is an equation-oriented flowsheeting tool for the design, simulation and optimization of chemical processes (Eich-Soellner *et al.*, 1997) with a long history of applications (Burr, 1991,/; Voith, 1991; Zapp and Sendler, 1993; Engl and Schmidt, 1996/06/26; Engl *et al.*, 1999; Kröner, 2006/07/09-13).

Dynamic model equations of unit operation which occur in LNG liquefaction processes, *i.e.*, mixes, splits, gas/liquid separators, throttle valves, compressors and two stream

[1]OPTISIM® is a registered trademark of the Linde AG.

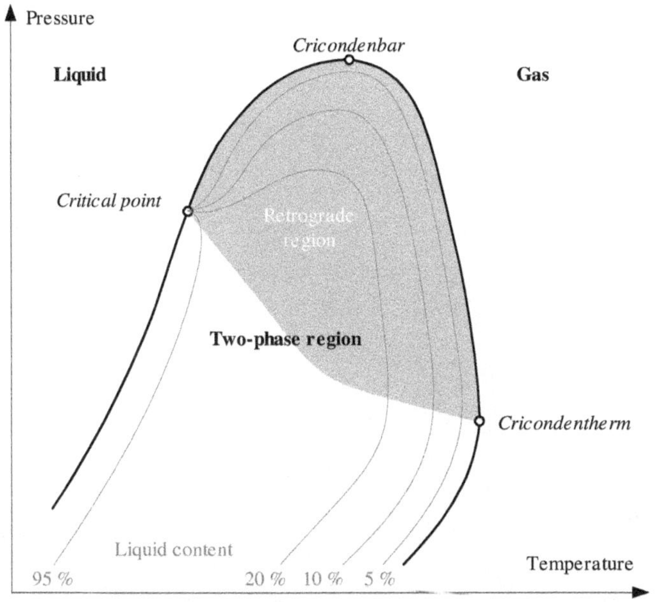

Figure 2.1.: Typical phase diagram of hydrocarbon mixture of fixed composition (from Hammer *et al.*, 2006, p. 7)

counter/cross-current heat exchangers are not presented here due to various illustrations in prior work (Melaaen, Oct 1994; Eich-Soellner *et al.*, 1997; Hangos and Cameron, 2001; Zaïm, Mar 2002; Hammer; Singh and Hovd, 2006/09/28-29). Deeper insights into the modeling of compressors are provided in the textbook of Lüdtke (2004). Within the scope of this thesis, the model equations of the spiral wound heat exchangers in the OPTISIM® simulator were revised. Therefore, a closer look at this unit operation is made in the next section.

2.4. Spiral-wound heat exchanger

This section deals with the development of a dynamical model of spiral-wound heat exchangers (SWHEs) for the Linde in-house simulator OPTISIM®. Properties of SWHEs and fields of application are introduced in Section 2.4.1. Related work in experimental investigation and dynamical modeling of SWHEs are referred to in Sections 2.4.2 and 2.4.3, respectively. Modeling issues are discussed in Section 2.4.4. In Section 2.4.5 and 2.4.6, governing conservative and phenomenology equations are presented. The dynamical validation of the resulting SWHE model takes place in Section 2.4.7. Concluding remarks are given in Section 2.4.8.

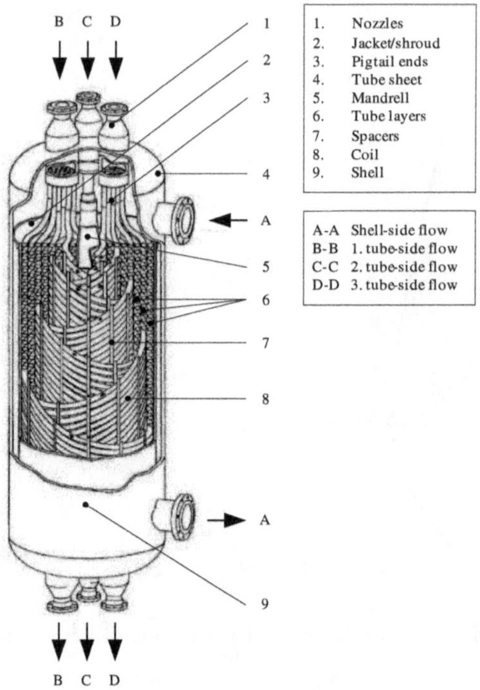

Figure 2.2.: Principal sketch of an SWHE (from Hausen and Linde, 1985, p. 472)

2.4.1. Introduction

In LNG liquefaction plants, natural gas is cooled against a coolant which is usually a mixture of light hydrocarbons. The cooling of natural gas takes place in so called main cryogenic heat exchangers (MCHE). Temperatures of approx. $-160\,°\mathrm{C}$ are sufficient to transform natural gas into its liquid state at standard pressure. Usually, spiral/coil-wound heat exchangers (SWHE/CWHE) are deployed as MCHEs.

Remark 2.1. For some reasons, plate-fin heat exchangers (PFHE) serve sometimes as a proper substitute. It is known that SWHEs are capable of large temperature differences and gradients. PFHEs instead require smooth operation, but are preferable when limited operation space is provided (see Linde publication "Looking inside ...", 2005). LNG plants built in the 60s and early 70s (based on different cycles) possessed both PFHE and SWHE equipment. Later, the SWHE became more and more dominant.

SWHEs are fairly complex units as they consist of up to 1000 km of coiled tubes in a shell. Abadzic and Scholz (1973) state that they offer unique advantages whenever (i) simultaneous heat transfer between more than two streams is desired, (ii) a large number of heat transfer units are required, and (iii) high operating pressures of the streams are given. SWHE can be used generally for all purposes where clean service is present, but are primarily used for low-temperature processes as feed coolers and liquefiers. In the scope of plants manufactured by the Engineering Division of the Linde AG, SWHEs are predominantly applied in LNG plants but not exclusively. Bach *et al.* (2001/05/14-17) state that from 1973 until 1992 SWHEs, installed by Linde were in operation in an air separation plant in Ludwigshafen, Germany (BASF). Other fields of applications are Ethylene plants, Rectisol units and CO shift conversion (listed in the Linde publication "Coil-wound heat exchanger", 2006a). Further fields of applications are illustrated in Thier and Backhaus (1997, p. 264)

The geometry of SWHEs can be varied widely to obtain optimal flow conditions for all streams and still meet heat transfer and pressure drop requirements. Figure 2.2 provides a simple sectional drawing of an SWHE. Tubes are wound around a core cylinder (the mandrel), which is mainly designated for stability during manufacturing, and collected in headers at both ends of the cylindrical shell. Successive layers of tubes separated by spacing strips are wound in opposite directions.

2.4.2. Experimental investigation

Neeraas (Sep 1993) investigated the tube-side heat transfer and pressure drop of pure hydrocarbons (C3, R22) and hydrocarbon mixtures (C1/C2, C2/C3) in SWHEs with a designated test rig (concentric and inclined double tube). He investigated two flow patterns, single phase and annular (*i.e.*, shear controlled) flow. The gravity controlled slug flow regime was not considered. Various correlations for single and two phase heat transfer and frictional pressure drop were validated. Sufficiently validated models are listed in Table 2.1. For heat transfer and friction in single phase flow, a correlation by Dittus-Boelter (1930)[2] and Gnielinski (1986a), respectively was successfully validated. For pure and mixed refrigerants at the annular flow state, good agreement between measured and predicted heat transfer and friction was obtained for the method by Boyko and Kruzhilin (1967) and Fuchs (Jul 1975), respectively. It is important to stress that a major functional dependency within the model by Fuchs (Jul 1975) is represented by a (non-physical) black-box model, *i.e.*, a curve fit of measurement data. This makes the method somewhat unreliable in terms of generality. The second best validated method for frictional pressure drop at annular flow state is the one by Friedel (1980). In order to account for secondary effects, Neeraas proposed correction methods for these empirical correlations. The method by Silver (1947), independently developed by Bell and Ghaly (1972), corrects the heat transfer coefficients by Boyko and Kruzhilin (1967) by taking the film thickness into account. A further enhancement of the heat transfer coefficients is

[2]presented by Shah and Sekulić (2003, pp. 482, 484)

Flow regime	Heat transfer	Frictional pressure drop
Single phase	Dittus-Boelter (1930)	Gnielinski (1986a)
Annular flow	Boyko and Kruzhilin (1967) (corr. by Silver, 1947 and Sardesai *et al.*, 1982, 1983)	Fuchs (Jul 1975)/ Friedel (1980)

Table 2.1.: Models for tube-side heat transfer and frictional pressure drop validated by Neeraas (Sep 1993)

related to mass transfer effects, significant in the presence of multi-component mixtures, which can be compensated for with the correction of Sardesai *et al.* (1982, 1983).

Equivalently to Neeraas (Sep 1993), Fredheim (May 1994) investigated the heat transfer and pressure drop of pure propane and nitrogen coolants, as well as ethane/propane mixtures on the shell-side of an SWHE in a test facility. He pointed out that different flow regimes are usually present along the path of the shell stream and introduced subsequent zones (from inflow at the top towards outflow at the bottom):

A Gravity drained environment, with a liquid film on the wall and low-vapor velocity in the annular space between the tubes

B Transient environment where both gravity force and vapor-shear force contribute to the liquid flow

C Shear-controlled environment, with a high vapor velocity, which enhances the fluid flow and the entrainment rate

D Superheated vapor flow

Accordingly, a transformation from a purely gravity controlled to shear controlled flow occurs along the path due to the increase of void fraction from usually $0.02 - 0.07\,\%$ at the top, causing the gas flow to accelerate considerably and entrain liquid droplets. For each flow regime, publicly available empirical correlations were fitted to measurement data obtained. Fredheim (May 1994, p. 90) pointed out that heat transfer of saturated falling film flow is complex due to four (main) mechanisms involved:

- Heat transfer by gravity-drained film flow
- Heat transfer by enhancement due to shear flow
- Heat transfer by nucleate boiling[3]
- Heat transfer reduction due to mixture effects

[3]In the state of nucleate boiling, vapor bubbles are generated over cavities on the hot surface remarkably affecting the heat transfer (Fredheim, May 1994, p. 75).

Flow regime	Heat transfer	Frictional pressure drop
Superheated	Gnielinski (1979)$^{\dagger\ddagger\wr}$; Abadzic (1974/11/17-22)‡	Barbe *et al.* (1972a)$^{a\dagger\ddagger\wr}$ Gilli (1965)$^{\dagger\ddagger\wr}$ (calculation of flow area)
Falling film	Bays and Mcadams (1937)$^{\dagger\ddagger\wr}$; Bennett *et al.* (1986)$^{\dagger\ddagger\wr}$	
Shear	McNaught (1982)†‡	Barbe *et al.* (1972) corrected based on suggestions of Grant and Chisholm (1979)†‡
Nucleate boiling	Stephan and Abdelsalam (1980)‡	

[a] C. Barbe, D. Mordillat, D. Roger. Pertes de charge en ecoulement monophasique et diphasique dans la calandre des exhangeurs bobins. *XII Journees de l'Hydraulique*, Paris, France, 1972.

Table 2.2.: Models for shell-side heat transfer and frictional pressure drop validated by Fredheim (May 1994, pp. 56-101)† , Aunan (2000)‡ and Neeraas *et al.* (2004a,b)$^{\wr}$

The models in Table 2.2 marked with † have been sufficiently validated by Fredheim (May 1994, pp. 56-101). The work of Fredheim has been continued by Aunan (2000) with an improved test facility. Sufficiently validated models are included in Table 2.2 and are indicated by ‡. Recently validated models by Neeraas *et al.* (2004a,b) are supplemented in Table 2.2 and marked with ≀. Note that later validation results are generally favored due to their better accuracy. The calculation of the free flow area can be done by the use of the method by Gilli (1965).

2.4.3. Dynamic modeling

For the purpose of dynamical LNG plant simulation, SWHE modeling was performed by Melaaen (Oct 1994), Zaïm (Mar 2002) and Hammer and Singh and Hovd (2006/09/28-29, 2007/05/27-30). The resulting models are based on the almost identical assumptions and simplifications with minor differences, *e.g.*, calculation methods for heat transfer coefficients and pressure drops. The most important modeling assumptions are given in the sequel.

- Application of a one-dimensional homogeneous flow model for both tube-side and shell-side streams.

- Consideration of a stationary mass balance, resulting in infinite composition dynamics.
- Consideration of enthalpy as a conservation state instead of internal energy (except for Hammer which compared both variants).
- Assumption of thermodynamic equilibrium between gas/liquid phases.
- Negligence of gravity (by Zaïm, Mar 2002; Hammer), heat radiation and axial conduction (fluid/wall).
- No heat conduction resistance of the tube wall in radial direction.
- Assumption of a lumped insulation heat flow through the shell wall instead of modeling natural convection in the jacket clearance. Consideration of an adiabatic mandrel (no heat accumulation).

SWHE model validation was performed by Hammer *et al.* (2003/03/30-04/03) and Vist *et al.* (2003/03/30-04/03)[4].

Remark 2.2. An unconventional dynamic SWHE model was recently proposed by Hasan *et al.* (2007/05/27-30). They modeled SWHEs by the use of heat exchanger networks. This approach requires no geometrical data of SWHEs and is thus convenient for black-box modeling. Drawbacks thereof are that model parameters need to be estimated for the SWHE, which requires transient measurement data of external states, *i.e.*, input/output, and that internal states of the heat exchanger network are not necessarily representative for the internal states of the SWHE.

2.4.4. Modeling issues

Modeling issues occur mainly because phase change is present on the shell-side and in particular tubes of the SWHE. In LNG applications, the shell-side fluid is evaporating and moving from top to bottom of the bundle while the tube-side streams typically flow upwards and are either cooled or condensed.

A brief survey of the field of gas/liquid flow is provided by Perry *et al.* (1999, pp. 6.26-6.29). The detailed modeling of gas/liquid flow is challenging due to physical complexity. Note that each phase has its own composition, density, viscosity and velocity spatially distributed. Another dimension of complexity is the fact that different flow regimes are

[4]They considered a slightly modified model of Hammer named DCOIL.

Figure 2.3.: Gas/liquid flow patterns in horizontal pipes (from Perry *et al.*, 1999, p. 6.26)

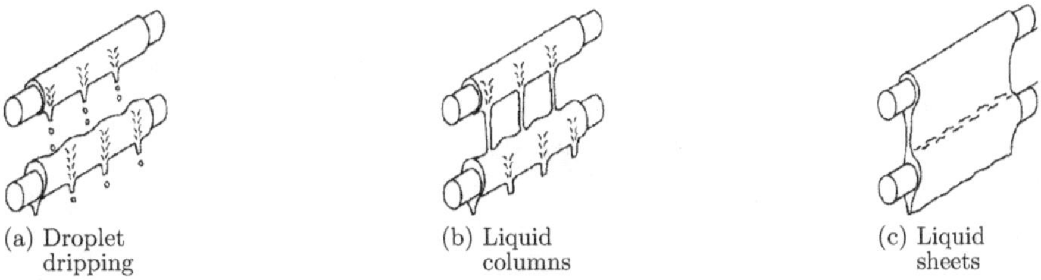

Figure 2.4.: Patterns of liquid flow between two adjacent horizontal tubes (from Kocamustafaogullari and Chen, 1988)

usually present along the path of phase change. A categorization of tube-side and shell-side flow regimes is shown in Figure 2.3 and 2.4, respectively. Models for the temporally and/or spatially distributed properties of two-phase flow such as phase inversion point[5], degree of dispersion, slip ratio, void fraction, interfacial area, mass transfer and frictional pressure drop, are usually rare. Besides, the fact that each flow regime possesses its own model equations makes an implementation into a DAE solver difficult.

2.4.4.1. Classification of flow models

Wallis (1969) presents three different modeling approaches for two-phase flow (listed with decreasing degree of sophistication):

Separated flow model It is taken into account that both phases have different properties. Models of various degrees of complexity may be derived. The most sophisticated version requires separate conservation equations for each phase and their simultaneous solution. Model equations describing the interaction between the phases and the phases and the wall are additionally necessary. Fundamental characteristics of two-phase flow are the slip ratio $S = w_g/w_l$ and the void fraction $\varepsilon = V_g/V_l = A_g/A_l$. A relationship between both can be obtained by a linear

[5]This is the point where the continuous phase transforms into the disperse and vice versa.

combination of the continuity equation for each phase and is given by

$$\varepsilon = \left(1 + S\,\frac{\rho_{\mathrm{g}}}{\rho_{\mathrm{l}}}\left(\frac{1-\omega}{\omega}\right)\right)^{-1}.$$

Drift flux model This model is essentially a special case of the separated flow model. Attention is focused on the relative motion of the two phases rather than on the motion of the individual phases. Ishii and Hibiki (2006, p. 382) state that the use of the drift flux model is appropriate when the motions of the two phases are strongly coupled. This is especially valid when the relative motion is independent of the flow rates of each phase (Wallis, 1969, p. 89). The drift flux model is thus convenient for bubbly, slug and droplet flow, but not for annular flow. The volumetric flux of either phase relative to a surface moving at the volumetric average velocity $j = j_1 + j_2$ is known as the drift flux $j_{21} = -j_{12}$ and is proportional to the relative velocity $w_{12} = w_1 - w_2$ according to

$$j_{21} = (1-\varepsilon)\; j_2 - \varepsilon\, j_1 = \varepsilon\,(1-\varepsilon)\; w_{12}.$$

The drift flux may be seen analogous to the diffusion flux in the molecular diffusion of gases. In terms of the fluxes of each phase, the mean density is given by

$$\overline{\rho} = \frac{j_1\,\rho_1 + j_2\,\rho_2}{j} + (\rho_1 - \rho_2)\,\frac{j_{21}}{j}.$$

Mixture conservation equations may be derived as presented in Ishii and Hibiki (2006, pp. 382-383).

Homogeneous flow model In the theory of homogeneous flow, the phases are considered being well mixed and may thus be conveniently represented by a homogeneous phase. All properties of the two phases are averaged between the phases by using a certain weighting method in order to apply conservation equations to the pseudo continuous fluid. For some properties this may be sufficient, particularly when one phase is finely dispersed into the other. Then, slip factor $S = 1$ holds and the (average) velocity of both phases reads

$$\bar{w} = 4\,\frac{\dot{M}}{\rho\,\pi\,d_{\mathrm{h}}^2}.$$

The homogeneous density follows from linear vapor fraction weighting of the specific

Stream	p/p^{crit}	ρ_g/ρ_l	η_g/η_l
Natural gas	0.6	0.1	0.2
Tube-side mixed refrigerant	0.4	0.1	0.1
Shell-side mixed refrigerant	0.04	0.03	0.01

Table 2.3.: Natural gas and mixed refrigerant properties of the Mossel Bay plant

volumes of both phases and can be written in terms of the individual densities as

$$\bar{\rho} = \frac{\rho_g\,\rho_l}{\rho_g + \dot{\omega}_g\,(\rho_l - \rho_g)}.$$

The determination of a characteristic viscosity is more problematic. Three different formulas are commonly used in literature (cited by Wallis, 1969, p. 27):

$$\text{Mc Adams } et\ al.\ (1942)\ \bar{\eta} = \frac{\eta_g\,\eta_l}{\eta_g + \dot{\omega}_g\,(\eta_l - \eta_g)}$$

$$\text{Cicchitti } et\ al.\ (1960)\ \bar{\eta} = \dot{\omega}_g\,\eta_g + (1 - \dot{\omega}_g)\,\eta_l$$

$$\text{Dukler } et\ al.\ (1964)\ \bar{\eta} = \bar{\rho}\left(\dot{\omega}_g\,\frac{\eta_g}{\rho_g} + (1 - \dot{\omega}_g)\,\frac{\eta_l}{\rho_l}\right).$$

According to Kraume (2004, pp. 457-458), these averaging methods provide quite different results. However, it must be stressed that it is anyway demanding to reliably predict the viscosities of single phase mixtures and therefore, the error related to the averaging between the phases is not generally the most dominant.

2.4.4.2. Significance for LNG service

It is likely that the tube-side streams in LNG applications possess high pressures close to the critical point. This leads to small property differences between the gas and the liquid phase and averaging of properties as necessary for the homogeneous flow model is thus appropriate. Table 2.3 shows pressure, density and viscosity data of the natural gas and the mixed refrigerant in the Mossel Bay plant. The pressure ratios p/p^{crit} of the tube-side fluids are at least 0.4. This forces the density and dynamical viscosity ratios of the tube-side fluids to be not smaller than 0.1. Hence, the driving forces for phase separations are small and the homogeneous model is likely to work out well on the tube side. This was also indicated by the measurement results of Neeraas (Sep 1993). His validation of heat transfer correlations based on the homogeneous model expectedly showed that the model prediction error decreases with increasing pressure.

Remark 2.3. The natural gas shows the highest pressure ratio, which is due to the commonly known fact that the liquefaction of a high pressure feed is favorable in terms of efficiency (Durr *et al.*, 2008; Yates, 2002/10/13-16). Besides, natural gas is conventionally liquefied at pressures higher than 40 bar in order to reach acceptable levels of compressor shaft power requirements. Accordingly, a minimal ratio of $p/p^{\text{crit}} \approx 0.6$ for natural gas can be considered.

It is important to stress that the velocities of the tube-side fluids are usually quite large (an example is given in Table 2.4), which provides sheer stress and favors a homogeneous distribution of the disperse phase (bubbles/droplets).

On the shell side, completely different conditions are present. On the one hand, the shell-side pressures are smaller than the tube-side ones. On the other hand, the shell-side velocities are usually smaller at the inlet section of the bundle as the flow is predominantly gravity driven (Fredheim, May 1994, p. 57). Hammer, p. 90 also pointed out that the residence time on the shell-side is several times larger than that of the tubes which he specified with $8 - 10\,\text{s}$. It can be concluded that for the shell-side flow a separated flow model is thus better suited than a homogeneous one. This was verified by Hammer *et al.* (2003/03/30-04/03), which found that the homogeneous flow model neglecting gravity is not sufficient to describe two-phase shell-side flow properly.

2.4.5. Conservation equations

In order to provide sufficient mathematical treatment, all tubes and the shell are considered one-dimensional. That is, state variables and properties are allowed to change with longitudinal spatial coordinate and time. Taking into account that the tube slope is constant for each particular tube pass, the use of a unique spatial coordinate[6]

$$\zeta = \frac{z_{\text{shell}}}{L_{\text{shell}}} = \frac{z_i}{L_i} = \frac{z_i}{L_{\text{shell}}} \sin\left(\beta_i\right)$$

is convenient, where $0 < \beta_i \leq \frac{\pi}{2}$. The energy e is conserved and consists of internal and kinematic energy. It is written as

$$e = h + \frac{p}{\rho} + w^2.$$

Further model assumptions are:

- All properties are assumed to be homogeneously distributed across the cross-section of one passage.
- Diffusion and conduction terms are negligible.
- Heat radiation is neglected over the diffusive and convective transport.

[6]The index i indicates the ith tube pass.

- Heat transfer is treated as conversion terms in the conservation equations.
- Heat transfer to the jacket/shroud is referred to as a static insulation heat term and heat transfer to the mandrel is negligible.

2.4.5.1. Governing fluid equations

The general formulation of the mass conservation equation for the species j is known as

$$\underbrace{\frac{\partial}{\partial t}\rho_j}_{\text{accumulation}} + \underbrace{\frac{1}{L}\frac{\partial}{\partial \zeta}\left(\rho_j w\right)}_{\text{transportation}} = 0. \tag{2.1}$$

The general formulation of the momentum conservation equation is given by

$$\underbrace{\frac{\partial}{\partial t}(\rho w)}_{\text{accumulation}} + \underbrace{\frac{1}{L}\frac{\partial}{\partial \zeta}\left(\rho w^2\right)}_{\text{transportation/acceleration}} = \underbrace{-\frac{1}{L}\frac{\partial}{\partial \zeta}p}_{\text{pressure force}} + \underbrace{\frac{1}{L}\frac{\partial}{\partial \zeta}\sigma}_{\text{frictional force}} + \underbrace{\rho g \sin(\beta)}_{\text{gravitational force}}, \tag{2.2}$$

where $\rho = \sum_{j=1}^{n_c} \rho_j$. Note that from now on, homogeneous properties for two phase flow are not explicitly indicated anymore by the bar notation. Finally, the general formulation of the energy conservation equation reads

$$\underbrace{\frac{\partial}{\partial t}(\rho e)}_{\text{accumulation}} + \underbrace{\frac{1}{L}\frac{\partial}{\partial \zeta}(\rho e w)}_{\text{transportation}} = \underbrace{-w\frac{1}{L}\frac{\partial}{\partial \zeta}p}_{\text{pressure work}} + \underbrace{w\frac{1}{L}\frac{\partial}{\partial \zeta}\sigma}_{\text{frictional work}} + \underbrace{w \rho g \sin(\beta)}_{\text{potential energy}} + \underbrace{\dot{q}}_{\text{heat transfer}}. \tag{2.3}$$

These equations hold for single and multi-phase flow. In order to solve the conservation equations, model reduction is reasonable to avoid computational expensiveness. Nevertheless, it is required that the solution of a simplified model remains realistic in terms of the main influential factors. The magnitudes of all terms in (2.1) through (2.3) give an impression about appropriate negligence. A basis for an estimation of these magnitudes is served by data in Tables 2.4 and 2.7 which show geometrical and process data of the Mossel Bay subcooler.

2.4.5.2. Substantiation of model reductions

It is convenient to simplify the energy expression $e = h - p/\rho + w^2$. According to the data in Table 2.4, the contribution of Δh in Δe is larger than 99 % for both the tube-side and the shell-side fluid. Thus, the substitution $e = h$ is justified for steady-state model-

	Tube-side LMR		Shell-side LMR	
Conditions	Inlet	Outlet	Inlet	Outlet
Temperature (in K)	215	113	112	211
Pressure (in bar)	38.4	36.8	4.3	4.1
Density $\left(\text{in kg/m}^3\right)$	417	635	590	220
Velocity (in m/s)	1.9	0.8	0.2	3.0
Enthalpy (in kJ/kg)	-470	-736	-736	-266

Table 2.4.: Inlet and outlet process data of the Mossel Bay subcooler

ing. Hammer, pp. 137-138 investigated the divergence of the two modeling approaches, enthalpy and internal energy conversion, by comparison of dynamical simulation results of an LNG liquefaction sub-model of the MFC® process. It turned out that the models behave differently at time scales smaller than $\mathcal{O}\left(10^2\,\text{s}\right)$.

An estimation of the magnitudes of the modeling equation terms can be done by making them dimensionless. For this sake, reduced variables need to be defined, *e.g.*, $p_r = p/\hat{p}$ in terms of pressure. If a variable is spatially distributed, its reference value, *i.e.*, $\hat{p}$ in the example, is estimated by the use of the geometric mean between inlet and outlet. The coefficients obtained are listed in Table 2.5. All coefficients are dimensionless except for the ones of the accumulation terms. Consequently, the negligence of the accumulation term may only be substantiated quantitatively if time scales are taken into account. For the momentum conservation equation it can be concluded that the transport term may be negligible over the pressure, frictional and gravity force (provided an error of 1 ‰ is admissible). The same holds for the momentum accumulation term if considered time scales are way above $\mathcal{O}\left(10^{-1}\,\text{s}\right)$. In contrast, the pressure and frictional work as well as the potential energy are negligible over the energy transportation term. For both the mass and energy conservation, the accumulation can be neglected against transportation if time scales way above $\mathcal{O}\left(10^2\,\text{s}\right)$ are considered which can be verified by prior works (*e.g.*, Melaaen, Oct 1994; Mandler *et al.*, May 1998; Hammer; Singh and Hovd, 2006/09/28-29) which indicate that LNG liquefaction process transients are in time scales of approximately an hour.

The main discrepancy between the model simplifications proposed in this section and the model implementation done by other authors such as Melaaen (Oct 1994), Zaïm (Mar 2002) and Hammer is that all considered the mass conservation equation quasi-stationary. In section 2.4.7.3 it is shown that a quasi-stationary continuity equation together with a dynamical energy conservation equation is not recommended due to poor simulation results.

	Mass	Momentum	Energy
Accumulation	$L/\hat{w}$	$L/\hat{w}$	$L/\hat{w}$
	$\sim \{10^2, 10^1\}$ s	$\sim \{10^2, 10^1\}$ s	$\sim \{10^2, 10^1\}$ s
Transportation	1	1	1
Pressure force/work		$-\hat{p}/\hat{\rho}\,\hat{w}^2$	$-\hat{p}/\hat{\rho}\,\hat{h}$
		$\sim -\{10^4, 10^3\}$	$\sim -\{10^{-2}, 10^{-3}\}$
Frict. force/work		$\hat{p}/\hat{\rho}\,\hat{w}^2$	$\hat{p}/\hat{\rho}\,\hat{h}$
		$\sim \{10^4, 10^3\}$	$\sim \{10^{-2}, 10^{-3}\}$
Gravity/pot. energy		$g\,L\,\sin(\beta)\,/\hat{w}^2$	$g\,L\,\sin(\beta)\,/\hat{h}$
		$\sim \{10^3, 10^3\}$	$\sim \{10^{-4}, 10^{-4}\}$

Table 2.5.: Magnitudes of terms of the dimensionless conservation equations for {tube, shell} -side based on LMR data of the Mossel Bay subcooler (according to Tables 2.4 and 2.7)

2.4.5.3. Reduced PDAE model

According to the comments above, the fluid is modeled using the following set of model equations:

$$\frac{\partial \boldsymbol{c}}{\partial t} = -\frac{1}{L}\frac{\partial (w\,\boldsymbol{c})}{\partial \zeta} \tag{2.4a}$$

$$0 = -\frac{1}{L}\frac{\partial p}{\partial \zeta} - f_{\text{fric}}(w, \boldsymbol{c}) \pm \tilde{\boldsymbol{M}}^T \boldsymbol{c}\,g \tag{2.4b}$$

$$\frac{\partial h}{\partial t} = -\frac{1}{L}\frac{\partial (w\,h)}{\partial \zeta} - \sum_i^{n_{\text{m}}} f_{\text{ht}}(T, p, \boldsymbol{c}, w, T_i^{\text{m}})\,(T - T_i^{\text{m}}). \tag{2.4c}$$

Note that for the sake of simple notation, the index i for the stream has been omitted. Moreover, the density vector $\boldsymbol{\rho}$ is replaced by the molar concentration vector $\boldsymbol{c}$. Caloric and thermal equations of state serve as closure conditions to meet zero degree of freedom of the fluid equations. They are respectively given by the general formulations

$$0 = T - f_{\text{ceos}}(h, p, \boldsymbol{c}) \tag{2.5a}$$

$$0 = p - f_{\text{teos}}(T, \boldsymbol{c}). \tag{2.5b}$$

The metal around each fluid is modeled by a simple energy balance of heat transferred between the metal and the streams, in which the temperature is independent of the radial coordinate and the enthalpy is conserved. *I.e.*, heat conduction is infinitely fast in radial direction and neglected in longitudinal direction. This yields

$$\frac{\partial h^{\text{m}}}{\partial t} = \sum_i^{n_{\text{s}}} f_{\text{ht},i}(T_i, p_i, \boldsymbol{c}_i, w_i, T^{\text{m}})\,(T_i - T^{\text{m}}).$$

A general formulation of the caloric equation of state serve as a relation between metal enthalpy and temperature, *i.e.*,

$$0 = T^{\mathrm{m}} - f^{\mathrm{m}}_{\mathrm{ceos}}\left(h^{\mathrm{m}}\right). \tag{2.6}$$

The model equations (2.4a) through (2.6) represent a coupled set of algebraic as well as ordinary and partial differential equations. Owing to their special properties, the set-up and solution of these so-called partial differential-algebraic equations (PDAEs) need careful attention. According to Le Lann *et al.* (1998/07/15), PDAEs are difficult to solve because change in a single parameter or boundary condition may lead to a completely different behavior of the solution. As a consequence, numerical methods which perform well for PDEs may be totally unable to do so for PDAEs. Systematic structure analyses and solution approaches of PDAEs came up by the late 1990s (an overview is given by Eichler-Liebenow, 1999/06/18, pp. 47-48) and publications in this field still happen to continue (for instance Neumann and Pantelides, 2008).

Some noteworthy properties of PDAEs are:

- PDAEs are non hyperbolic due to the contribution of algebraic equations and equations with partial derivatives with respect to the spatial coordinate (Martinson and Barton, 2001).

- According to Lieberstein (1972, p. 73), a PDE system is said to be well-posed if it has a unique solution and depends continuously on its data. Similarly to PDEs, a PDAE system may be improperly stated (not well-posed), which makes it not amenable for numeric integration by standard integration codes. Martinson and Barton (2001) provide a framework which helps to identify improperly stated PDAEs.

- A PDAE system possesses two indices, each of them at least of number 1. According to Martinson and Barton (2000), they are called (differentiation) index with respect to t and x, respectively, and are defined as a natural generalization of the (differentiation) index of a DAE defined by Brenan *et al.* (1987, p. 17).

- By the method-of-lines, PDAEs can be transformed into DAEs. The resulting DAE may or may not be of high index.

A linear stability analysis of a two pass heat exchanger model similar to the PDAE system (2.4a) through (2.6) was performed by Hanke *et al.* (2005) with the difference that a pure fluid was considered. A generalization thereof is presented in Appendix 2.A. The analysis reveals that boundary conditions (BCs) must be imposed for the mass flux $\boldsymbol{c}\,|v|$ and the

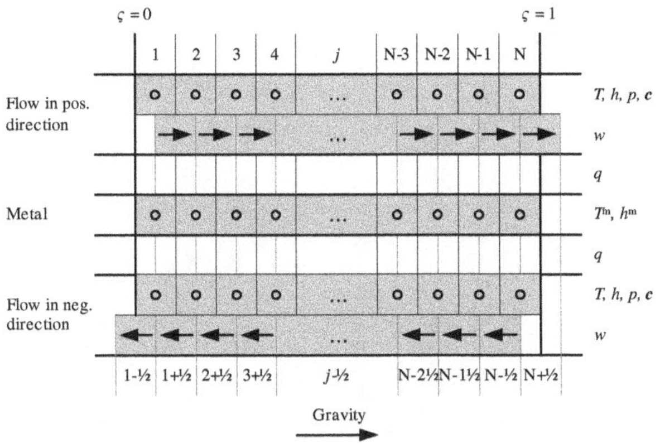

Figure 2.5.: Discretization scheme for a staggered velocity grid

enthalpy h at inlet and for the pressure p at outlet of the passage, *i.e.*,

$$
\begin{aligned}
0 &= \boldsymbol{c}\left(\zeta_{\text{in}}\right) \left|w\left(\zeta_{\text{in}}\right)\right| A - \boldsymbol{F}_{\text{in}} \\
0 &= p\left(\zeta_{\text{out}}\right) - p_{\text{out}} \\
0 &= \left|w\left(\zeta_{\text{in}}\right)\right| h\left(\zeta_{\text{in}}\right) - \dot{H}_{\text{in}}.
\end{aligned}
$$

Unlike the BCs, the initial conditions are not arbitrary. Consistent initial conditions for p, h, $\boldsymbol{c}$ are needed and usually obtained by solving the stationary form of the PDAE.

2.4.5.4. Transformation into DAEs

The method-of-lines is applied for the transformation of the PDAE into its DAE form. It refers to the discretization of the spatial differentials by using the backward form of the first order finite difference approximation. Such (upwind) methods are appropriate as they have proven to be stable provided that their stability conditions are satisfied. A frequently applied first order upwind discretization method is the method by Godunov. In combination with an explicit Euler integration algorithm, the stability is proven if $v \leq 1$ where $v = \left|\bar{\lambda}\right| \Delta t / \Delta x$ is the Courant number and $\bar{\lambda}$ is the largest wave speed, *i.e.*, the maximum eigenvalue of the coefficient matrix of the hyperbolic system (see for instance LeVeque, 2002, p. 70).

The discretization applied is based on the discretization scheme shown in Figure 2.5. The relating semi-discrete equations for the fluids read

	$w > 0$	$w < 0$
j	$\in \{2, \ldots, N\}$	$\in \{1, \ldots, N-1\}$
j_{u}	$j-1$	$j+1$
$j_{\frac{1}{2}}$	$j+1/2$	$j-1/2$
$j_{\frac{1}{2}\mathrm{u}}$	$j-1/2$	$j+1/2$
j_{m}	j	j

Table 2.6.: Index notation for discretization

$$
\begin{aligned}
\frac{\partial \boldsymbol{c}_j}{\partial t} &= -\frac{1}{L}\frac{1}{\Delta\zeta}\left(\left|w_{j_{\frac{1}{2}}}\right| \boldsymbol{c}_j - \left|w_{j_{\frac{1}{2}\mathrm{u}}}\right| \boldsymbol{c}_{j_{\mathrm{u}}}\right) \\
0 &= -\frac{1}{L}\frac{p_j - p_{j_{\mathrm{u}}}}{\Delta\zeta} - f_{\mathrm{fric}}\left(w_{j_{\frac{1}{2}\mathrm{u}}}, \boldsymbol{c}_{j_{\mathrm{u}}}\right) + \mathrm{sgn}\left(w_{j_{\frac{1}{2}\mathrm{u}}}\right) \tilde{\boldsymbol{M}}^T \boldsymbol{c}_{j_{\mathrm{u}}} g \\
\frac{\partial h_j}{\partial t} &= -\frac{1}{L}\frac{1}{\Delta\zeta}\left(\left|w_{j_{\frac{1}{2}}}\right| h_j - \left|w_{j_{\frac{1}{2}\mathrm{u}}}\right| h_{j_{\mathrm{u}}}\right) - \sum_i^{n_{\mathrm{m}}} f_{\mathrm{ht},i}\left(T_j, p_j, \boldsymbol{c}_{,j}, w_{j_{\frac{1}{2}}}, T_{j_{\mathrm{m}}}^{\mathrm{m},i}\right)\left(T_j - T_{j_{\mathrm{m}}}^{\mathrm{m},i}\right) \\
0 &= f_{\mathrm{ceos}}\left(h_j, T_j, p_j, \boldsymbol{c}_j\right) \\
0 &= f_{\mathrm{teos}}\left(T_j, p_j, \boldsymbol{c}_j\right),
\end{aligned}
$$

with the index notation indicated in Table 2.6. The discretization of the metal energy conservation equation yields

$$
\begin{aligned}
\frac{\partial h_{j_{\mathrm{m}}}^{\mathrm{m}}}{\partial t} &= \sum_i^{n_{\mathrm{s}}} f_{\mathrm{ht,i}}\left(T_j^i, p_j^i, \boldsymbol{c}_j^i, w_{j_{\frac{1}{2}}}^i, T_{j_{\mathrm{m}}}^{\mathrm{m}}\right)\left(T_j^i - T_{j_{\mathrm{m}}}^{\mathrm{m}}\right) \\
0 &= f_{\mathrm{ceos}}^{\mathrm{m}}\left(h_{j_{\mathrm{m}}}^{\mathrm{m}}, T_{j_{\mathrm{m}}}^{\mathrm{m}}\right).
\end{aligned}
$$

The BCs read

$$
\begin{aligned}
0 &= \boldsymbol{c}_{j_{\mathrm{in}}} \left|w_{j_{\mathrm{in}}}\right| A - \boldsymbol{F}_{\mathrm{in}} \\
0 &= p_{j_{\mathrm{out}}} - p_{\mathrm{out}} \\
0 &= \begin{cases} w_{j_{\mathrm{in}}} h_{j_{\mathrm{in}}} - \dot{H}_{\mathrm{in}} & \text{pure} \\ T_{j_{\mathrm{in}}} - T_{\mathrm{in}} & \text{otherwise} \end{cases},
\end{aligned}
$$

where $j_{\mathrm{in}} = \begin{cases} 1 & \text{if } w > 0 \\ N & \text{otherwise} \end{cases}$ and $j_{\mathrm{out}} = \begin{cases} N & \text{if } w > 0 \\ 1 & \text{otherwise} \end{cases}$.

2.4.6. Phenomenology

In the equations above, heat transfer and pressure drop correlations are generally referred to as f_{ht} and f_{fric}, respectively. In this section, they are specified in more detail. As pointed out in Section 2.4.2, different stream regimes are present along the passage on both tube and shell side which relates to a considerable amount of empirical correlations needed. Modeling effort can be constrained by imposing restrictions on the generality and therefore reducing the number of necessary empirical correlations. When only LNG service is focused, the following restrictions can be imposed.

Tube-side condensation and shell-side evaporation Empirical models for tube-side evaporation and shell-side condensation can be disregarded.

Downward flow on shell side Only models for gravity driven falling film flow and shear flow need to be considered on the shell side besides the model for (superheated) single phase flow.

Small temperature differences The rather complicated empirical model for nucleate boiling can be disregarded for shell-side evaporation. Instead, the simpler tube-side condensation model serves as an equivalent substitute.

Selected models referred to in Tables 2.1 and 2.2 have been implemented into the OPTISIM® SWHE model. They are indicated in more detail in Appendix 2.B. In order to achieve good convergence performance of the model equations, the transitions between the correlations of two adjoined flow regimes have been smoothed (once continuously differentiable). Note that implementation into OPTISIM® requires the statement of derivatives with respect to the state variables, *i.e.*, p, $\mathbf{c}$, w and h in each cell. Due to the high complexity of the correlations, automatic differentiation technique (ADOL-C, Sep 2005) was chosen and successfully applied.

2.4.7. Dynamical validation

Operational data of the peakshaving LNG liquefaction plant Mossel Bay served as a basis for the validation of the OPTISIM® SWHE model.

2.4.7.1. Test plant set-up

A detailed description of the tests was presented by Reithmeier *et al.* (2004/04/21-24). Here, the most important facts are recalled. In 1998, an industrial sized SWHE prototype manufactured by Linde was installed in parallel to the MCHE of the Mossel Bay LNG liquefaction plant commissioned in 1992. The liquefaction process is based on a single mixed refrigerant (SMR) cycle. The mixed refrigerant consists mainly of nitrogen, methane, ethylene and isobutane. Both the SWHE and the test streams have been extensively instrumented. A total of 32 temperature elements were installed at the shell-side of the three bundles. Figure 2.6 shows the installation of one temperature element inside

Figure 2.6.: Installation of a PT-100 element inside an SWHE bundle (from Reithmeier *et al.*, 2004/04/21-24)

the bundle. All measurements were individually judged by redundancy in order to ensure high accuracy. Beside the "day to day" operation, designated tests were carried out in two successive test periods. The first test period took place from September through November 1999. Its focus was mainly the study of SWHE performance at steady state operation as well as plant start-up and trip tests. After three years of flawless production, a second test period was carried out from April through May 2001. Its objective was the investigation of the dynamic behavior of the SWHE as well as the confirmation of the original 100 % load. The dynamic investigations included the following scenarios.

- Feed gas trips, refrigerant compressor trips and plant trips[7]
- Plant shutdowns
- Plant start-ups from warm/cold conditions
- Cooling down/warming up procedures due to unbalanced supply and demand of refrigeration[8]
- Step changes of other process parameters at different loads

2.4.7.2. Modeling and data manipulation

The transient measurements from the second test period were used to validate the SWHE model. The simulation of the complete Mossel Bay liquefaction unit was not considered appropriate for the purpose of SWHE model validation. This is due to the fact that model inaccuracies of supplementary equipment such as compressors, compressor aftercoolers *etc.* would have to be taken into account. Instead the dynamic validation of one stand-alone SWHE bundle including closely related equipment was considered. Transient input, output and (partly) internal conditions for each bundle were available. The decision to

[7] These incidents either happened during normal operation due to different causes or were forced deliberately by tripping the refrigerant compressor.

[8] These scenarios were created by either a rapid increase/reduction in mixed refrigerant load at constant natural gas flow or a rapid reduction/increase of natural gas flow at constant mixed refrigerant load. The conditions were held constant within a certain time period after which the process parameters were set back to their original values.

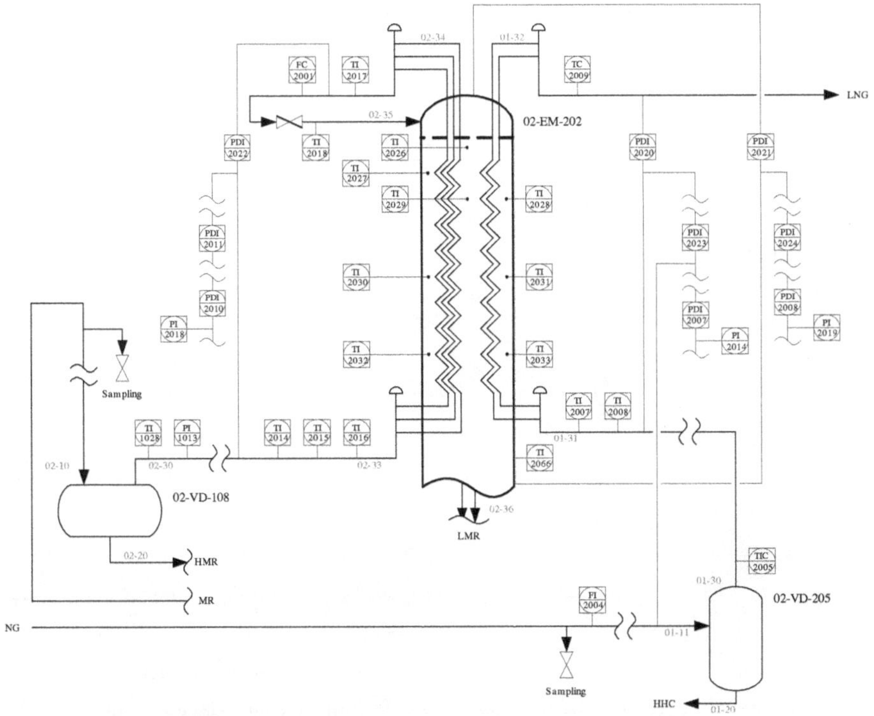

Figure 2.7.: Instrumentation of the subcooling bundle of the Mossel Bay SWHE

investigate the subcooling bundle 02-EM-202 was based on the facts that (i) the least complexity is present due to a minimum of two tube passes and that (ii) the interesting case of phase change is present in all three tube passes.

The instrumentation for data acquisition is given in Figure 2.7. The mixed refrigerant and natural gas composition were detected by manual sampling (laboratory analyzes) several times a day. In order to account for transient composition changes, the heavy hydrocarbons (HHC) separator and the mixed refrigerant separator, 02-VD-205 and 02-VD-108, respectively, were modeled supplementarily to the bundle. Accordingly, the temperatures and pressures inside the separators became additional transient input variables to be regarded. The data set consisting of measurements every 30 s was manually resampled into larger non-equidistant time steps in order to reduce noise and accelerate simulation time.

Geometrical data of the Mossel Bay LNG subcooler required for the calculation of shell/tube masses and stream velocities were obtained from design drawings and data sheets delivered by the Linde proprietary design tool GENIUS (Steinbauer and Hecht, 1996/06/03-05). In Table 2.7 the data is presented for convenience. The Mossel Bay

	Tubes		
	Natural gas	Mixed refr.	Shell
Outer diameter tube/shell (in mm)	12	12	1360
Tube/shell wall thickness (in mm)	1	1	26
Tube/shell length (in m)	101.4	118.6	9.6
Number of tubes	137	336	
Diameter mandrel (in m)			0.6
Thickness of radial spacers (in mm)			2.0
Mean tube distance (in mm)			2.7
Number of tube layers			27
Metal density $\left(\text{in } \frac{\text{kg}}{\text{m}^3}\right)$	2700	2700	2700
Metal mass (in Mg)	1.30	3.72	2.82
Net cross flow area $\left(\text{in m}^2\right)$	0.043	0.106	0.224

Table 2.7.: Geometrical data of the Mossel Bay SWHE

SWHE was made of aluminum.

Some comments regarding modeling issues are given in the sequel.

Remark 2.4. After modeling the OPTISIM® flowsheet, initial values for all input variables were set. Expectedly, owing to modeling errors, the calculated steady state model output did not agree with the measurement output. In order to fix the discrepancy, model parameters needed to be adjusted. Heat exchanger tuning factors, multipliers of heat transfer and frictional pressure drop coefficients, were chosen for this purpose. The number of independent tuning factors for fitting the output temperatures was equal to the number of heat exchange areas. Thus, as the heat exchange area between shell and ambient atmosphere was neglected, only two degrees of freedom were present to manipulate three output temperatures. Consequently, one further independent tuning parameter was necessary to fit the three calculated output temperatures to the measurement data. The strategy to correct the natural gas flow rate did not work out well because quite a considerable correction of approximately 30 % less natural gas flow rate would have to be necessary which would intolerably affect the heat exchanger dynamics. Therefore, the correction of the natural gas flow rate was omitted willing an offset of 1.5 K between calculated and measured output temperature of the mixed refrigerant stream at the warm end of the bundle. Note that a discrepancy at the warm end is more reasonable than at the cold due to maximum spatial temperature gradient at the warm end as indicated in Figure 2.8.

Remark 2.5. It turned out that it is inappropriate not to model the choke valve and consider the stream 02-35 as an independent dynamic input of the model. In this case, the model output at the cold end would show excellent agreement with measurement data despite significantly inaccurate internal conditions. This effect is related to very

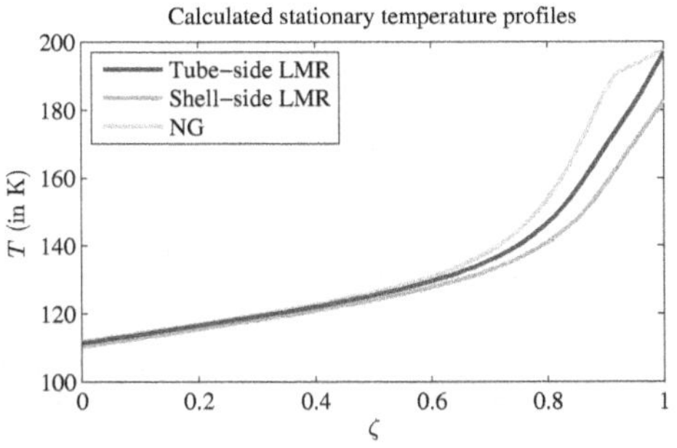

Figure 2.8.: Stationary temperature profiles in the Mossel Bay subcooler

small (driving) temperature differences at the cold end due to the large heat exchanger area. As a consequence, the choke valve was modeled as an isenthalpic expansion to a pressure given by measurement data.

2.4.7.3. Simulation runs

Dynamic simulations were carried out for two operation scenarios and four different sophistication levels of the modeling equations. The level of sophistication is indicated by a three digit flag code, referring whether the dynamical accumulation of the metal energy, the stream energy and the mass is considered (1) or not (0). *I.e.*, the code (0,0,0) relates to pure steady-state models for reference purposes, (1,0,0) indicates that metal energy is the only accumulator, (1,1,0) states infinite mass dynamics and finite metal/stream energy dynamics and (1,1,1) is the most sophisticated model where metal/stream energy and mass dynamics are finite. The respective numbers of differential equations per accumulating cell yield 0, 1, 2 and $2 + n_c$. The number of accumulating cells is given by

$$\#\text{passages} \times (\#\text{cells per passage} - 1)\,.$$

The SWHE model was set up with 50 cells and a mixed refrigerant with 10 components. Accordingly, the total numbers of differential equations for models of increasing sophistication are 0, 147, 294 and 1764. The simulation were conducted using a Windows XP® SP2 desktop with Intel® Core™ Duo CPU E8400 (3.0 GHz, 3.5 GB RAM).

The first validation run is based on the measurement data obtained from 7:15 to 9:15 PM on May 14th, 2001. The measurement data around the subcooler is shown in Figure 2.10. After steady state operation, the natural gas flow rate was ramped from 17200 to 14300 $\mathrm{Sm^3/h}$ at 7:35 PM. As a result, the flow rate of the light mixed refrigerant (LMR) was reduced by the controllers. This caused the discharge pressure to increase until at 8:07 PM (3120 s) the venting of mixed refrigerant was executed. Figure 2.10

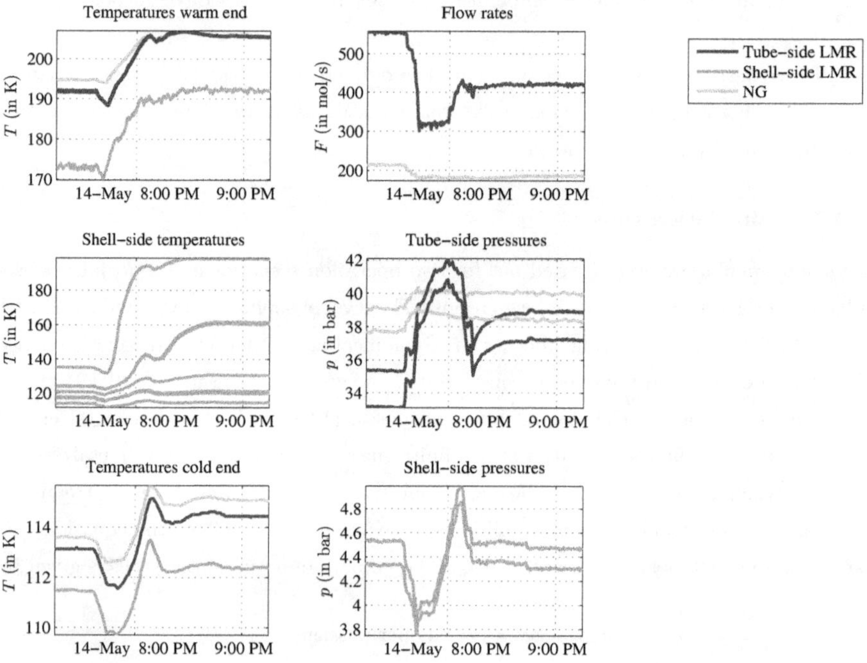

Figure 2.9.: Measurement data from May 14th, 2001

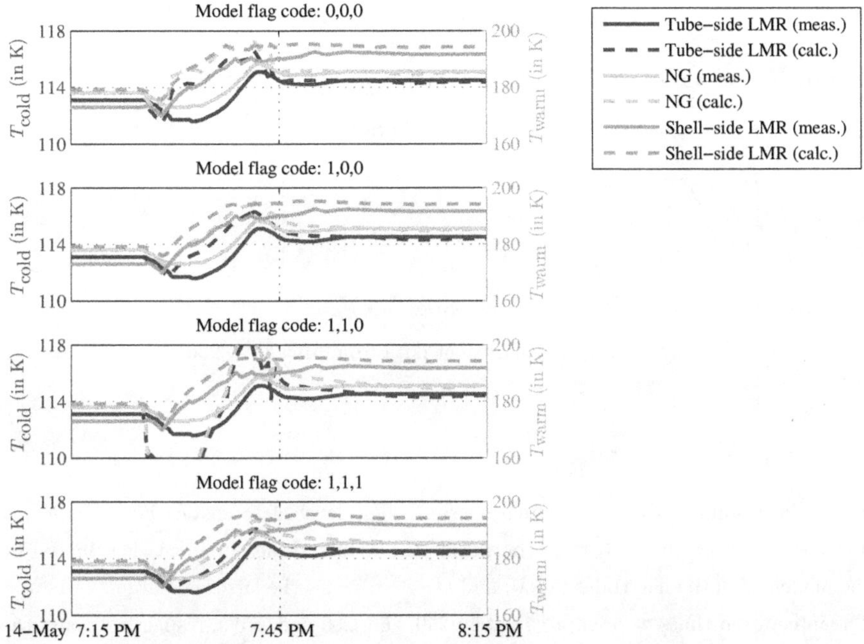

Figure 2.10.: Comparison between measurement data from May 14th, 2001 and simulation results

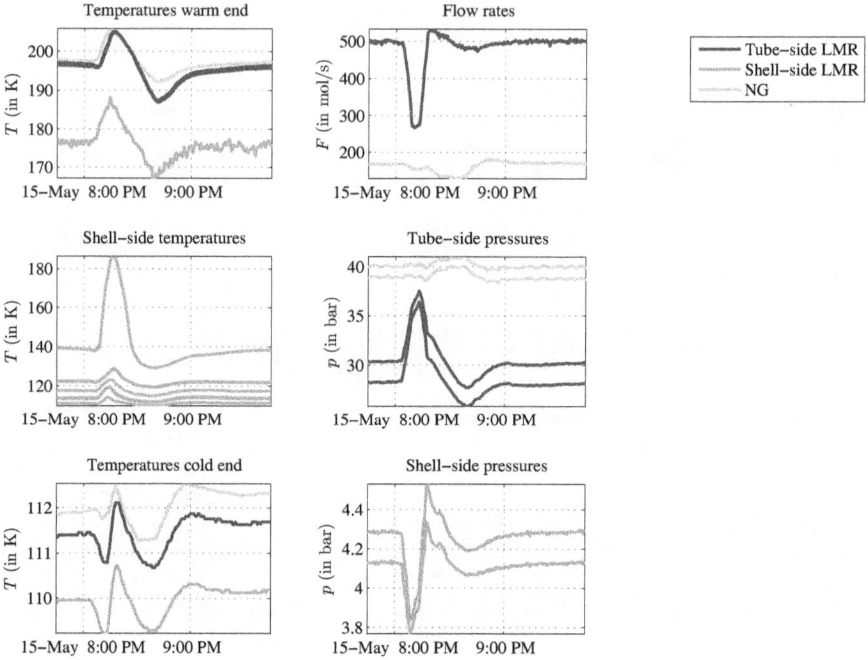

Figure 2.11.: Measurement data from May 15th, 2001

shows the results of the four simulation runs. The level of sophistication is indicated above each chart. The best match between simulation and measurements is achieved by the models with the flag codes (1,0,0) and (1,1,1) whereas the latter is marginally better. The integration times were approximately 240, 260, 200 and 270 s, respectively, and the number of integration steps were in the order of 1300 for all. Interesting to note is the fact that the model with flag code (1,1,0) produced the worst predictions despite its rather high level of sophistication.

The measurement set from 7:15 to 9:15 PM on May 15th, 2001 was selected for a second validation run. The data is shown in Figure 2.11. The test can be seen as a kind of impulse step response. At 8:03 PM, both the HMR and LMR flow rate were reduced by 45 % and shortly afterwards at 8:12 PM the prior conditions were restored. At 9:30 PM the plant returned into stable operation. Figure 2.12 shows the simulation results. A similar picture as for the first validation run can be drawn. The number of integration steps and the integration times are in the same order of magnitude.

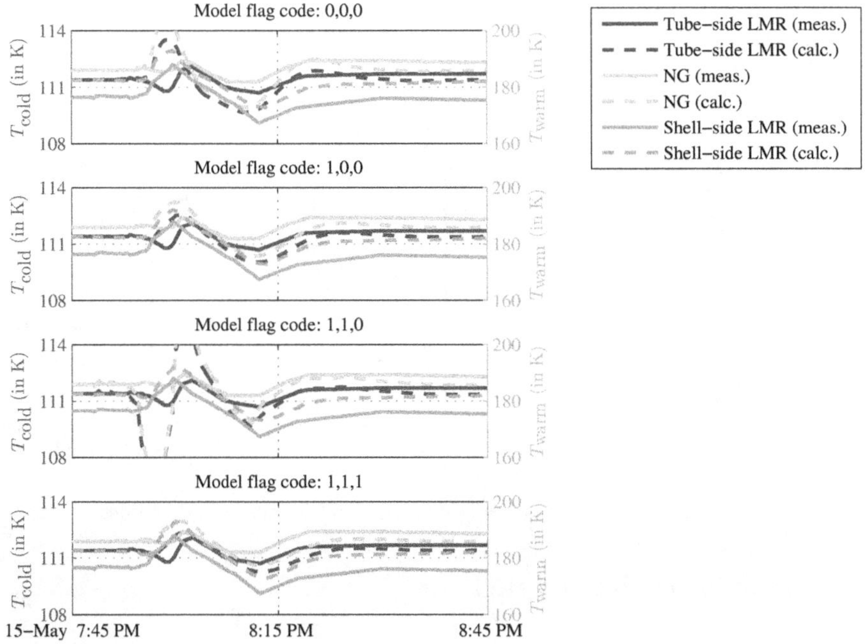

Figure 2.12.: Comparison between measurement data from May 15th, 2001 and simulation results

2.4.8. Conclusions

In this section, the development of a dynamical model of an SWHE unit for the use in the Linde in-house simulator OPTISIM® is presented. In contrast to related dynamical models in the open literature, the model allows for temporal mass accumulation. Besides, the model includes empirical correlations of heat transfer and pressure drop specially validated for SWHEs applied in LNG liquefaction plants. The SWHE model has limited capabilities as it is based on a homogeneous fluid model. *I.e.*, scenarios where gas and liquid phase have considerable different velocities cannot be accurately predicted. For instance, the effect that liquid gathers at the bottom of the SWHE on both tube and shell side during trips or shut down scenarios cannot be calculated. Further enhancement of the SWHE model may thus include a separated flow model predominantly for the shell-side. For better predictions on smaller time scales, stream internal energy conservation could be implemented (Hammer).

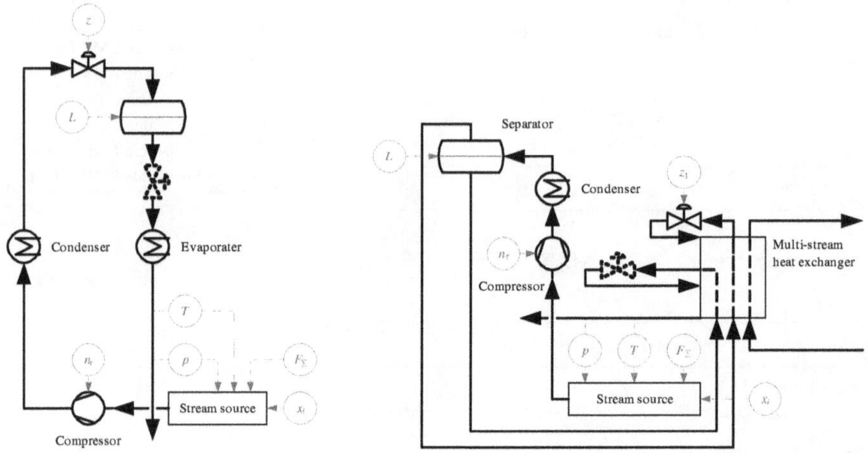

(a) Refrigeration cycle with variable active charge

(b) Refrigeration cycle with variable active charge and partly condensation

Figure 2.13.: Partly closed configurations of refrigeration cycle models

2.5. Refrigeration cycles

In this section some issues related to the modeling of cycle processes are discussed.

2.5.1. Steady-state models

Steady state models of mixed refrigerant cycles are built as shown in Figure 2.13a and b. Both cycles are modeled with an inlet and an outlet stream and are closed by assigning the temperature and the pressure of the outlet stream to the inlet stream. This is called *partly closed configuration*. Note that neither of the cycles can be directly closed by connecting the outlet to the inlet stream as this would yield a singular set of equations, *i.e.*, the flow rate and composition would be undetermined. In each cycle, one flow controlling valve (dotted) is inactive for the sake of simplicity. *I.e.*, the valve is modeled as a constant pressure drop, as its degree of freedom is moved to the cycle inlet flow rate which can be independently adjusted instead.

At first glance, this partly closed configuration seems unproblematic and it is indeed if the inlet flow rate and inlet composition are both independent of process conditions. However, independence of inlet flow rate gets lost if the active charge becomes fixed as in certain cycle topologies discussed by Jensen and Skogestad (2007b) and independence of inlet composition gets lost if partly condensation of mixed refrigerants takes place.

Loss of independence of inlet flow rate through fixed charge can be proven by considering the cycle in Figure 2.13a. Here, the fixed active charge can be implemented by either omitting the liquid storage tank after the throttle valve or by keeping the level in the tank fixed. Without loss of generality it is assumed that a pure refrigerant is in duty and that the major mass reservoirs are the liquid in the condenser, the evaporator and in between (piping and tank). On the one hand, the level depends on ambient conditions, *i.e.*, the temperature of the heat exchanging mediums in the condenser and evaporator as they influence the density of the liquid refrigerant and therefore the liquid volume. On the other hand, the level depends on the flow rate of the refrigerant as it determines both the condensation and evaporation front individually. Now, it can be concluded that if the level in the tank is fixed (or the tank is omitted), then the flow rate of the refrigerant must depend on ambient conditions and cannot be independent anymore. Thus, one equation must be added in the modeling equations in which the volume integral over the refrigerant density equals a fixed value. The corresponding additional (dependent) variable is the cycle inlet flow rate.

Loss of independence of the refrigerant composition in case of partly condensation can be shown by considering the cycle shown in Figure 2.13b. Without loss of generality it can be supposed that the heavy mixed refrigerant (HMR) liquid in the separator is the only reservoir in the cycle. If the temperature of the heat exchanging medium in the condenser drops, the HMR becomes lighter but cannot as the separator liquid is the only reservoir. Instead the separator inlet stream must become heavier in order to leave the HMR composition invariant. Accordingly, the mixed refrigerant composition at the cycle inlet is dependent on ambient conditions. The appropriate modeling of the mixed refrigerant cycle is performed by introducing n_c equations where n_c is the number of species in the refrigerant. In the ith equation the volume integral of the molar density of the ith species over all equipment is fixed to a constant value. The $n_c - 1$ inlet molar fractions and the separator level are introduced as (dependent) variables. This can then be referred to as a *quasi-closed configuration*.

2.5.2. Dynamic models

Unlike as in steady-state models the cycle outlet stream can be directly connected to the cycle inlet stream. Dynamic solution (*i.e.*, integration of the DAE system) can take place after calculating the initial solution and switching the model equations from partly closed or quasi-closed configuration to an actually closed configuration. In order to get an appropriate initial steady-state solution, the model setup must nevertheless be realized as discussed in the prior section.

The dynamic simulation of refrigeration cycles involves yet another modeling issue. As mentioned in Section 2.4.5.3, the appropriate boundary conditions of distributed heat exchangers models depend on the configuration of the model equations, *i.e.*, a well posed

PDE is required. If only the energy conservation equations of the heat exchanging streams are dynamic, it is sufficient to determine all state variables at the inlet boundary of each passage. These boundary conditions fail if the energy and mass conservation equations are dynamic and the momentum conservation equation is quasi-stationary. In this case it is necessary that the flow of each species and the temperature are imposed at the inlet and the pressure is imposed at the outlet of each passage. If such heat exchanger models are used for modeling refrigeration cycles, then there are in fact two loops which need to be built. The first loop consists of stream connectors and carries the flow rate, composition and temperature information. The second loop consists of variable connectors and propagates the pressure information in countercurrent to the first loop. It is important to stress that for modeling pressure driven flow this is anyway recommended in order to avoid generating a high index DAE.

2.A. Stability analysis of the fluid model

Based on the demonstration of Hanke *et al.* (2007), the necessary conditions for the stability of the PDAE subsystem (2.4a) through (2.5b) are presented in this section. It is convenient to combine the thermal and caloric EOSs, (2.5a) and (2.5b), to one correlation of the form $h = g\left(p, \boldsymbol{c}\right)$ by elimination of the temperature and consider a reduced set of three primary variables $\boldsymbol{c}$, p and $H = w\, g$. The reformulation of the PDAE in terms of these variables yields

$$g\, \boldsymbol{c}_t + H\, \boldsymbol{c}_x + \boldsymbol{c}\, H \left(\frac{H_x}{H} - \frac{g_p}{g}\, p_x - \frac{g_{\boldsymbol{c}}^T}{g}\, \boldsymbol{c}_x \right) = 0 \tag{2.7a}$$

$$p_x = S\left(H, p, \boldsymbol{c}\right) \tag{2.7b}$$

$$g_{\boldsymbol{c}}^T\, \boldsymbol{c}_t + g_p\, p_t + H_x = T\left(H, p, \boldsymbol{c}\right). \tag{2.7c}$$

Here, S and T represent momentum loss by friction/gravity and heat exchange, respectively. The indices t and x indicate temporal and spatial derivatives, respectively. The highly nonlinear PDAE (2.7a) through (2.7c) need to be linearized and freezed at a point of interest in order to conclude to local stability. With $\boldsymbol{u} = \begin{bmatrix} \boldsymbol{c}^T & p & H \end{bmatrix}^T$ the quasi-linear system can be stated as

$$\boldsymbol{A}\, \boldsymbol{u}_t + \boldsymbol{B}\, \boldsymbol{u}_x + \boldsymbol{C}\, \boldsymbol{u} = \boldsymbol{v}\left(x, t\right), \tag{2.8}$$

where

$$\boldsymbol{A} = \begin{bmatrix} \hat{g}\boldsymbol{I} & \boldsymbol{0} & \boldsymbol{0} \\ \boldsymbol{0} & 0 & 0 \\ \hat{g}_{\boldsymbol{c}}^T & \hat{g}_p & 0 \end{bmatrix}$$

$$\boldsymbol{B} = \begin{bmatrix} \hat{H}\boldsymbol{I} - \frac{\hat{H}}{\hat{g}}\hat{\boldsymbol{c}}\,\hat{g}_{\boldsymbol{c}}^T & -\frac{\hat{H}\,\hat{g}_p}{\hat{g}}\hat{\boldsymbol{c}} & \hat{\boldsymbol{c}} \\ 0 & 1 & 0 \\ 0 & 0 & 1 \end{bmatrix}.$$

Note that the freezed properties are indicated by the hat. Since the matrix pencil $\{\boldsymbol{A}, \boldsymbol{B}\}$ is regular[9], the system (2.8) may be transformed into its Weierstrass canonical form as suggested by Shirvani and So (1998). Transformation matrices $\boldsymbol{P}$ and $\boldsymbol{Q}$ have been derived using the symbolic computation functionality of MATHEMATICA® v5.2. The left-side multiplication of (2.8) with $\boldsymbol{P}$ yields

$$\begin{bmatrix} \hat{w}^{-1}\boldsymbol{I}_{n_c} & & \\ & 0 & \\ & -\hat{w}^{-1} & 0 \end{bmatrix} \boldsymbol{Q}^{-1}\boldsymbol{u}_t + \boldsymbol{I}\,\boldsymbol{Q}^{-1}\boldsymbol{u}_x + \boldsymbol{P}\,\boldsymbol{C}\,\boldsymbol{Q}\,\boldsymbol{Q}^{-1}\boldsymbol{u} = \boldsymbol{P}\,\boldsymbol{v}\,(x,t)\,, \tag{2.9}$$

where

$$\boldsymbol{Q}^{-1} = \begin{bmatrix} \hat{g}_{\boldsymbol{c}}^T & -\hat{g}_p\,\frac{1}{\hat{g}/(\hat{\boldsymbol{c}}^T\,\hat{g}_{\boldsymbol{c}})+1} & \\ \boldsymbol{D}_1 & \hat{g}_p\,\frac{1}{\hat{g}/(\hat{\boldsymbol{c}}^T\,\hat{g}_{\boldsymbol{c}})+1}\,\boldsymbol{1}_{n_c} & \\ -\hat{g}_{\boldsymbol{c}}^T & \hat{g}_p\,\frac{1}{\hat{g}/(\hat{\boldsymbol{c}}^T\,\hat{g}_{\boldsymbol{c}})+1} & \hat{w}^{-1} \end{bmatrix}, \tag{2.10a}$$

$$\boldsymbol{P} = \frac{\hat{g}}{\left(\hat{g} + \hat{\boldsymbol{c}}^T\,\hat{g}_{\boldsymbol{c}}\right)\,\hat{H}} \begin{bmatrix} -\hat{g}_{\boldsymbol{c}}^T & & \hat{\boldsymbol{c}}^T\,\hat{g}_{\boldsymbol{c}} \\ -\boldsymbol{1}_{n_c-1}\,\hat{g}_{\boldsymbol{c}}^T/\hat{g} + \boldsymbol{D}_2 & & -\boldsymbol{1}_{n_c-1} \\ & -\hat{g}_p\,\hat{H} & \\ \hat{g}_{\boldsymbol{c}}^T & & \hat{g} \end{bmatrix} \tag{2.10b}$$

[9]For information regarding matrix pencils and generalized eigensystems the reader is referred to matrix computation textbooks such as the one by Golub and VanLoan (1996, pp. 375-376).

in which

$$\boldsymbol{D}_1 = \begin{bmatrix} & & 1/\hat{c}_{n_c} \\ & \cdot^{\cdot^{\cdot}} & \\ 1/\hat{c}_2 & & \\ 0 & & \end{bmatrix} \in \mathbb{R}^{n_c \times n_c}$$

$$\boldsymbol{D}_2 = \frac{\hat{g} + \hat{\boldsymbol{c}}^T \hat{g}_c}{\hat{g}} \begin{bmatrix} & & 1/\hat{\boldsymbol{c}}_{n_c} \\ & \cdot^{\cdot^{\cdot}} & \\ 0 \quad 1/\hat{\boldsymbol{c}}_2 & & \end{bmatrix} \in \mathbb{R}^{(n_c-1) \times n_c}$$

and $\mathbf{1}_i = \begin{bmatrix} 1 & \dots & 1 \end{bmatrix}^T \in \mathbb{R}^i$. According to Martinson and Barton (2001), the first n_c rows of (2.9) are known as the hyperbolic subsystem, whereas the last two are referred to as the parabolic part. The advantage of the present formulation is that system properties become obvious by inspection and conclusions can be easily drawn. Generally, the following can be stated:

- The matrix $\boldsymbol{Q}^{-1}$ transforms the primary variables into variables known as invariants. Each subsystem (hyperbolic/parabolic) is equivalent to a system of ordinary differential equations along a particular direction in the (t, x) plane (Martinson, Feb 2000, pp. 100-106). The direction is given by the generalized eigenvalues $\boldsymbol{\lambda}(\boldsymbol{A}, \boldsymbol{B})$ that corresponds to the invariants of the sub-block.

- For the hyperbolic part, the boundary conditions on the left and right side are equal to the number of associated positive and negative generalized eigenvalues. The number of initial conditions equals the number the dimension of the hyperbolic part. If any generalized eigenvalue has nonzero degeneracy, the system does not depend continuously on its data.

- The number of boundary conditions required for the parabolic part equals its dimensions. Initial values are needed and restrictions on the side of the boundary conditions arise when an associated generalized eigenvalue is degenerated. More precisely, if a generalized eigenvalue is one-fold degenerated then it must be satisfied that (i) one initial condition is specified, (ii) the associated boundary conditions are not enforced at the same point and (iii) the index with respect to time of the entire system is less than two. With the latter two the solution depends continuously on its data.

From these facts the following conclusions can be drawn for (2.9).

- The number of n_c initial conditions must be specified for $\left(\boldsymbol{Q}^{-1}\boldsymbol{u}\right)_i \; \forall i \in \{1, \dots, n_c\}$.

- Additionally n_c BC for $\left(\boldsymbol{Q}^{-1}\boldsymbol{u}\right)_i$ $\forall i \in \{1,\ldots,n_c\}$ need to be imposed at inflow; depending on the velocity direction at the left ($w > 0$) or right ($w < 0$) boundary.

- The eigenvalues associated with the parabolic block are one-fold degenerated. Consequently, one initial condition is needed for $\left(\boldsymbol{Q}^{-1}\boldsymbol{u}\right)_{n_c+1} \propto p$. The lack of an initial condition for H is reasonable since it follows from $g\left(p,\boldsymbol{c}\right)$ and the respective initial conditions.

- Due to the degeneracy of the parabolic block, the BCs for $\left(\boldsymbol{Q}^{-1}\boldsymbol{u}\right)_{n_c+1}$ and $\left(\boldsymbol{Q}^{-1}\boldsymbol{u}\right)_{n_c+2}$ must be specified opposite at either side of the domain. From a practical point of view it is clear that the enthalpy is specified at inflow. Consequently, $\left(\boldsymbol{Q}^{-1}\boldsymbol{u}\right)_{n_c+2} = f\left(\boldsymbol{c},p,H\right)$ need to be imposed at inflow and $\left(\boldsymbol{Q}^{-1}\boldsymbol{u}\right)_{n_c+1} \propto p$ at outflow. Accordingly, the pressure must be specified at outflow in order to obtain a well-posed problem. Unfortunately, one further necessary condition for the well-posedness of the system can not be generally proven. As stated above, the index with respect to time must be less than two. However, it can only be generally shown that the upper bound for the index with respect to time of a linear system such as (2.10a) is less equal two[10].

In this section, the necessary BCs for the stable solution of a special class of heat exchanger models were derived. However, this knowledge is not new as it was previously pointed out by other authors (Pantelides, Feb 1998; Hanke *et al.*, 2007).

2.B. Empirical correlations

The tube arrangement within an SWHE can be specified with geometrical variables as indicated in the sectional drawing in Figure 2.14. Glaser (1938) and Abadzic and Scholz (1973) indicate how the mean gap width and thus the net cross-flow area among the tubes can be calculated. The gap width between two adjoining tubes is a function of the (constant) thickness of the spacers a, the (constant) height between the windings c and the (variable) relative position of the tubes x. Two characteristic positions may be determined. First, $x = 0$ where the gap width becomes minimal, *i.e.*, $s_{\max} = s_{\min} = a$. Second, $x = P_l$ where the gap width becomes maximal, *i.e.*, $s_{\max} = \sqrt{\frac{1}{4}P_l^2 + P_r^2} - d_o$. The mean gap width can expressed as

[10]Martinson and Barton (2002, Theorem 4.1) proved that the upper bound for the index with respect to time of a linear system with linear forcing is less equal the nilpotency of the coefficient matrix of the parabolic part. Note that $\begin{bmatrix} 0 & 0 \\ k & 0 \end{bmatrix}^2 = 0$.

Flow regime	**Tube side/single phase flow**
Author	Gnielinski (1986a)
Dim.less num.	$Re = \frac{w \rho d}{\eta}$
Validity	$1 < Re \sqrt{\frac{d}{D}} < 10^5$
Model equ.	$f_{\text{fric}} = f\left(Re, \frac{d}{D}\right) \frac{1}{d} \frac{\rho w^2}{2}$ $f\left(Re, \frac{d}{D}\right) = \begin{cases} \frac{64}{Re}\left(1 + 0.033\left(\log_{10}\left(Re\sqrt{\frac{d}{D}}\right)\right)\right) & \text{if } 1 < Re\sqrt{\frac{d}{D}} < Re_{\text{crit}} \\ \left(\frac{0.3164}{Re^{0.25}} + 0.03\sqrt{\frac{d}{D}}\right) & \text{if } Re_{\text{crit}} < Re\sqrt{\frac{d}{D}} < 10^5 \end{cases}$ $Re_{\text{crit}} = 2300\left(1 + 8.6\left(\frac{d}{D}\right)^{0.45}\right)$
Comments	–

Flow regime	**Tube side/annular flow**
Author	Chisholm (1973)
Dim.less num.	$X_{\text{tt}} = \left(\frac{\rho_{\text{l}}}{\rho_{\text{g}}}\right)\left(\frac{\rho_{\text{g}}}{\rho_{\text{l}}}\right)^{\frac{n}{2}}$
Validity	–
Model equ.	$f_{\text{fric}}^{\text{2ph}} = f_{\text{fric}}^{\text{l}}\left(1 + \left(X_{\text{tt}}^2 - 1\right)\left(B\left(x\left(1-x\right)\right)^{\frac{2-n}{2}} + x^{2-n}\right)\right)$ $B = \begin{cases} 4.8 & \text{if } X_{\text{tt}} \leq 9.5 \wedge G \leq 500 \\ \frac{2400}{G} & \text{if } X_{\text{tt}} \leq 9.5 \wedge 500 < G < 1900 \\ \frac{55}{G^{0.8}} & \text{if } X_{\text{tt}} \leq 9.5 \wedge G \geq 1900 \\ \frac{520}{\sqrt{X_{\text{tt}} G}} & \text{if } 9.5 < X_{\text{tt}} < 28 \wedge G \leq 600 \\ \frac{21}{X_{\text{tt}}} & \text{if } 9.5 < X_{\text{tt}} < 28 \wedge G > 600 \\ \frac{15000}{X_{\text{tt}}\sqrt{G}} & \text{if } X_{\text{tt}} > 9.5 \end{cases}$
Comments	The variable n refers to the exponent of Re in the Blasius equation; *i.e.*, $n = -\frac{1}{4}$. The variable G is the mass flux. For calculation of B, the unit of G must be $\text{kg}/\left(\text{s}\,\text{m}^2\right)$.

Table 2.8.: Tube-side pressure drop correlations

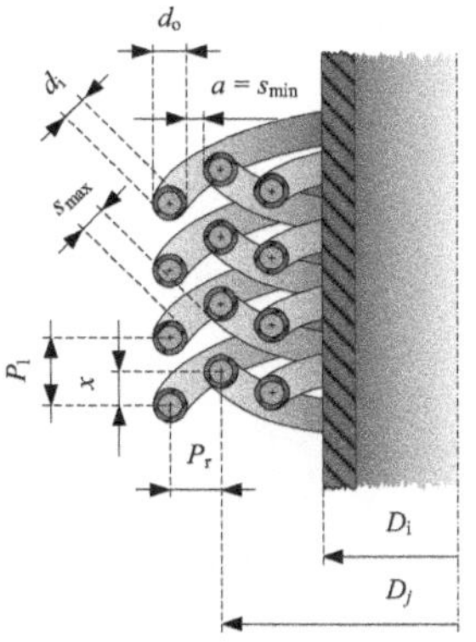

Figure 2.14.: Geometric characterization of the tube arrangement of an SWHE (from Abadzic and Scholz, 1973)

Flow regime	**Tube side/single phase flow**
Author	Dittus-Boelter (1930)
Dim.less num.	$Re = \frac{w\, d_i\, \rho}{\eta}$, $Pr = \frac{\eta\, c_p}{\lambda}$, $Nu = \frac{\alpha\, d_i}{\lambda}$
Validity	$2.5 \cdot 10^3 < Re < 1.24 \cdot 10^5$, $0.7 < Pr < 120$ (Shah and Sekulić, 2003, pp. 482, 484)
Model equ.	$Nu_{st} = 0.023\, Re^{0.8}\, Pr^{0.4}$ Curvature correction in turbulent flow: $\frac{Nu_{ct}}{Nu_{st}} = 1 + k\,\frac{d_i}{D}$, with $k \approx 3.5$
Comments	The subscript st and ct refer to *straight tube* and *coiled tube*, respectively. If curvature is small, the contribution of secondary flow is negligible and straight tube equations are adequate (Kreis, 1997).
Flow regime	**Tube side/annular flow**
Author	Boyko and Kruzhilin (1967), Silver (1947)
Dim.less num.	Re_l,Pr_l,Nu_l, $Re_g = \frac{(w_g - w_l)\,(d_i - 2\,\delta)\,\rho_g}{\eta_g}$, Pr_g, Nu_g
Validity	$1.5 \cdot 10^3 < Re < 1.5 \cdot 10^4$
Model equ.	$\alpha_{2ph} = \frac{\alpha_{film}}{1 + \alpha_{film}\,\Phi}$ (method by Silver, 1947; Bell and Ghaly, 1972) $\alpha_{film} = \alpha_l \sqrt{\frac{\rho_l}{\bar{\rho}}} = \alpha_l \sqrt{1 + \frac{\rho_l - \rho_g}{\rho_g}\,\varepsilon}$ (method by Boyko and Kruzhilin, 1967) $\Phi = \frac{Z}{\alpha_g\, C_f\, \theta}$ $Z = \frac{q_{core}}{q_{tot}} = \varepsilon\, (c_p)_g\, \frac{d_i}{d_i - 2\,\delta}\, \frac{\partial T}{\partial h}$ $\alpha_g = \frac{d_i}{d_i - 2\,\delta}\, Nu_{ct}\, \frac{\lambda}{d}$
Comments	The term $\frac{d}{d - 2\,\delta}$ may be disregarded if the film thickness is not taken into account. Not considered are (i) enhancement multiplier for the effect of liquid film turbulence C_f and (ii) the correction for mass transfer $\theta = \frac{\phi}{e^{\phi} - 1}$ where ϕ is the Ackermann correction term (as defined by Sardesai *et al.*, 1982, 1983).

Table 2.9.: Tube-side heat transfer correlations

$$s_{\mathrm{m}} = \frac{2}{P_{\mathrm{l}}} \int_0^{P_{\mathrm{l}}/2} s \, \mathrm{d}x$$
$$= \frac{P_{\mathrm{r}}}{2} \sqrt{1 + \frac{1}{4} \left(\frac{P_{\mathrm{l}}}{P_{\mathrm{r}}} \right)^2} + \frac{P_{\mathrm{r}}^2}{P_{\mathrm{l}}} \ln \left(\frac{P_{\mathrm{l}}}{2\, P_{\mathrm{r}}} + \sqrt{1 + \frac{1}{4} \left(\frac{P_{\mathrm{l}}}{P_{\mathrm{r}}} \right)^2} \right) - d_{\mathrm{o}}$$

and

$$s_{\mathrm{m}} = 1.04\, a + 0.04\, d_{\mathrm{o}}$$

for the special case $P_{\mathrm{l}} = P_{\mathrm{r}}$. From this result, the net cross flow area yields

$$A_f = \pi \, D_{\mathrm{m}} \, n_{\mathrm{lay}} \, s_{\mathrm{m}},$$

where

$$D_{\mathrm{m}} = D_{\mathrm{i}} + \left(n_{\mathrm{lay}} - 1 \right) \, a + n_{\mathrm{lay}} \, d + s_{\mathrm{m}} = \frac{D_{\mathrm{i}} + D_{\mathrm{o}}}{2}.$$

For more geometrical considerations, the reader is referred to Fredheim (May 1994, Appendix A).

Based on the geometrical variables in the sectional drawing of Figure 2.14, empirical correlations for tube/shell-side heat transfer and pressure drop are given in Tables 2.8 through 2.11 for various flow regimes. Depending on the context, the variable d refers to d_{o} or d_{i}.

Bibliography

Looking inside ...: Spiral-wound versus plate-fin heat exchangers. Pullach, Germany, 2005.

Coil-wound heat exchanger. Pullach, Germany, 2006a.

ADOL-C: A package for automatic differentiation of algorithms written in C/C++: v1.10.1, Sep 2005. URL `http://www.math.tu-dresden.de/wir/project/adolc/`.

E. E. Abadzic. Heat transfer on coiled tubular matrix: 74-WA/HT-64. The American Society of Mechanical Engineers winter annual meeting, New York, New York, 1974/11/17-22.

E. E. Abadzic and H. W. Scholz. Coiled tubular heat exchangers. *Advances in cryogenic engineering*, 18:42–51, 1973.

B. Aunan. *"Shell side heat transfer and pressure drop in coil-wound LNG heat exchangers:*

Flow regime	**Shell side/superheated flow**
Author	Barbe *et al.* (1972a) (for reference see Table 2.2)
Dim.less num.	$Re = \frac{d_o \rho w}{\eta} a = \frac{P_r}{d_o}$, $b = \frac{P_l}{d_o}$
Validity	$Re < 10^5$
Model equ.	$f_{\text{fric}} = 4 \left(\frac{1}{\sqrt{f_{st}(Re)}} + \frac{1}{\sqrt{f_{in}(Re)}} \right)^{-2} \frac{M^2}{2 \rho P_l}$
	$f_{st}(Re) = 0.88 \left(\frac{2a-1}{\sqrt{a^2+0.25 b^2}} + 1 \right)^2 \left(\frac{2(a-1)}{2a-1} \right)^{1.73} Re^{-0.295}$
	$f_{in}(Re) =$
	$\begin{cases} 1.52 (a-1)^{-0.7} (b-1)^{0.2} Re^{0.2} & \text{if } P_r \leq P_l \\ 0.32 \left(\frac{a-1}{b-1} - 0.9 \right)^{-0.68} (b-1)^{-0.5} Re^{0.2 \left(\frac{a-1}{b-1} \right)^2} & \text{otherwise} \end{cases}$
Comments	The functions f_{st} and f_{in} respectively represent the contribution of staggered and inline heat exchanger configuration.
Flow regime	**Shell side/shear flow**
Author	Chisholm (1973)
Dim.less num.	$X_{tt} = \left(\frac{\rho_l}{\rho_g} \right) \left(\frac{\rho_g}{\rho_l} \right)^{\frac{n}{2}}$
Validity	–
Model equ.	$f_{\text{fric}}^{2ph} = f_{\text{fric}}^{l} \left(1 + \left(X_{tt}^2 - 1 \right) \left(B \left(x (1-x) \right)^{\frac{2-n}{2}} + x^{2-n} \right) \right)$
Comments	See tube-side annular flow for definition of B and n.

Table 2.10.: Shell-side pressure drop correlations

Laboratory measurements and modelling: Ph.D. thesis. Ph.D. thesis, NTNU, Trondheim, Norway, 2000.

W. A. Bach, W. Förg, M. Steinbauer, R. Stockmann, and F. Voggenreiter. Spiral wound heat exchangers for LNG baseload plants. LNG conference, Seoul, South Korea, 2001/05/14-17.

C. Barbe, D. Roger, and D. Grange. Two-phase flow heat exchanges and pressure losses in spool-wound exchanger shells. *Pipeline and gas journal*, (11):82–87, 1972.

P. I. Barton. The equation oriented strategy for process flowsheeting. Cambridge, Massachusetts, Mar 2000.

G. S. Bays and W. H. Mcadams. Heat transfer coefficients in falling film heater: Streamline flow. *Industrial and engineering chemistry*, 29(11):1240–1246, 1937.

K. J. Bell and M. A. Ghaly. An approximate generalized design method for multicomponent/partial condensers. *AIChE symposium series*, 69:72–79, 1972.

D. L. Bennett, B. L. Hertzler, and C. E. Kalb. Down-flow shell-side forced convective boiling. *American institute of chemical engineers journal*, 32(12):1963–1970, 1986.

Flow regime	**Shell side/falling film**
Author	Bays and Mcadams (1937); Bennett *et al.* (1986)
Dim.less num.	$Re_\mathrm{l} = 4\,\frac{\Gamma}{\mu_\mathrm{l}}$, $Pr_\mathrm{l} = \frac{(c_p)_\mathrm{l}\,\mu_\mathrm{l}}{\lambda_\mathrm{l}}$ with $\Gamma = \frac{\dot{M}_\mathrm{l}}{2X}$ and $X = \pi\left(\frac{D_\mathrm{core}+D_\mathrm{shell}}{2}\right) n_\mathrm{layer}$
Validity	$40 < Re < 10^3$ (Bennett *et al.*, 1986), $2 \cdot 10^3 < Re < 10^4$ (Neeraas *et al.*, 2004b)
Model equ.	$Nu_\mathrm{l} = a\left(\frac{d_o}{\delta_\mathrm{c}}\right)^c Re_\mathrm{l}^b\, Pr_\mathrm{l}^b$ $\delta_\mathrm{c} = \left(\frac{\mu_\mathrm{l}^2}{g\,\rho_\mathrm{l}^2}\right)^{1/3}$
Comments	Sufficiently validated coefficients are $\{a,b,c,d\} =$ $\begin{cases}\{0.886, \frac{1}{9}, -\frac{1}{3}, \frac{1}{3}\} & \text{if } Re \le 2000 \text{ (Bennet et al. , 1986)} \\ \{0.313, \frac{1}{4}, -\frac{1}{3}, \frac{1}{3}\} & \text{otherwise (Neeraas et al. , 2004)}\end{cases}$
Flow regime	**Shell side/shear flow**
Author	McNaught (1982)
Dim.less num.	Lockhardt-Martinelli parameter $X_\mathrm{tt} = \left(\frac{1-\dot{x}}{\dot{x}}\right)^{0.9}\left(\frac{\rho_\mathrm{g}}{\rho_\mathrm{l}}\right)^{0.5}\left(\frac{\mu_\mathrm{l}}{\mu_\mathrm{g}}\right)^{0.1}$
Validity	$10^{-2} < X_\mathrm{tt} < 10^{-1}$
Model equ.	$\left(\alpha_\mathrm{2ph}\right)_\mathrm{s} = a\left(\frac{1}{X_\mathrm{tt}}\right)^b \alpha_\mathrm{ls}$ with $a = 1.26$ and $b = 0.78$
Comments	The subscript s indicates forced convective shear flow. The subscript ls indicates the case in which liquid is flowing alone in the section.
Flow regime	**Shell side/superheated flow**
Author	Gnielinski (1979)
Dim.less num.	$Re = \frac{\pi}{2}\frac{\rho\, w\, d_o}{\mu\,\gamma}$, $Nu = \frac{\pi}{2}\frac{\alpha\, d_o}{\lambda}$, $Pr = \frac{c_p\,\mu}{\lambda}$
Validity range	$1 < Re < 10^6$, $0.7 < Pr < 0.7 \cdot 10^3$
Model equ.	$Nu = f\left(\frac{P_\mathrm{l}}{P_\mathrm{r}}, \epsilon\right)\left(0.3 + \sqrt{Nu_\mathrm{lam}^2 + Nu_\mathrm{turb}^2}\right)$ $Nu_\mathrm{lam} = 0.664\sqrt{Re}\,Pr^{1/3}$ $Nu_\mathrm{turb} = \frac{0.037\,Re^{0.8}\,Pr}{1+2.443\,Re^{-0.1}\,(Pr^{2/3}-1)}$ $f\left(\frac{P_\mathrm{l}}{P_\mathrm{r}}, \epsilon\right) = 1 + \frac{0.7\left(\frac{P_\mathrm{l}}{P_\mathrm{r}}-0.3\right)}{\epsilon^{1.5}\left(\frac{P_\mathrm{l}}{P_\mathrm{r}}+0.7\right)^2}$
Comments	The variable $\epsilon = 1 - \frac{\pi}{4}\frac{d_o}{P_\mathrm{r}}$ represents the void fraction for calculation of average velocity between tubes.

Table 2.11.: Shell-side heat transfer correlations

L. D. Boyko and G. N. Kruzhilin. Heat transfer and hydraulic resistance during condensation of steam in a horizontal tube and in a bundle of tubes. *International journal of heat and mass transfer*, 10:361–373, 1967.

K. E. Brenan, S. V. La Campbell, and L. R. Petzold. *Numerical solution of initial-value problems in differential-algebraic equations*, volume 14 of *Classics in applied mathematics*. SIAM, Philadelphia, Pennsylvania, 1987. ISBN 0898713536.

P. S. Burr. Multi-phase problems and their solution with the 'MPFPL' simulator. In *Institution of Chemical Engineers symposia series*, volume 92, pages 329–340. 1985.

P. S. Burr. On-line optimization of an olefin plant complex with the OPTISIM equation-oriented simulator. *Oil gas european magazine*, 4:32–33, 1991.

P. S. Burr. The design of optimal air separation and liquefaction processes with the OPTISIM equation-oriented simulator, and its application to online and off-line plant optimization. AIChE spring national meeting, Houston, Texas, 1991/4/07-11.

P. S. Burr and A. M. Pfeiffer. MPFPL: A new Flowsheeting Program with Powerful Multi-Phase Capabilities. International congress of refrigeration, Paris, France, 1983.

D. Chisholm. Pressure gradients due to friction during the flow of evaporating two-phase mixtures in smooth tubes and channels. *International journal of heat and mass transfer*, 16(Feb):347–358, 1973.

C. Durr, C. Caswell, and H. Kotzot. LNG technology: The next chapter. *Hydrocarbon processing*, 87(7):39–56, 2008.

E. Eich-Soellner, P. Lory, P. S. Burr, and A. Kröner. Stationary and dynamical flowsheeting in the chemical industry. *Surveys on mathematics for industry*, 7:1–28, 1997.

C. Eichler-Liebenow. *Zur numerischen Behandlung räumlich mehrdimensionaler parabolischer Differentialgleichungen mit linear-implizierten Splitting-Methoden und linearer partieller differentiell-algebraischer Systeme: Ph.D. thesis.* Ph.D. thesis, Martin-Luther-Universität, Halle-Wittenberg, Germany, 1999/06/18.

G. Engl and A. Kröner. Success factors for CAPE in the engineering practice of a process plant contractor. Symposium on Process Systems Engineering/European Symposium on Computer Aided Process Engineering, Garmisch-Partenkirchen, Germany, 2006/07/09-13.

G. Engl and H. Schmidt. The optimization of natural gas liquefaction processes. The European Consortium for Mathematics in Industry, Lyngby, Denmark, 1996/06/26.

G. Engl, A. Kröner, T. Kronseder, and O. Styrk. Numerical simulation and optimal control of air separation plants. In H.-J. Bungartz, F. Durst, and C. Zenger, editors, *High performance scientific and engineering computing: Proceedings of the International FORTWIHR Conference on HPSEC*, volume 8 of *Lecture notes in computational science and engineering*, pages 221–231. Springer, 1999. ISBN 3540657304.

A. O. Fredheim. *Thermal design of coil-wound LNG heat exchangers: Shell-side heat transfer and pressure drop: Ph.D. thesis.* Ph.D. thesis, NTH, Trondheim, Norway, May 1994.

L. Friedel. Pressure drop during gas/vapor-liquid flow in pipes. *International chemical engineering*, 20(3):352–367, 1980.

P. H. Fuchs. *Trykkfall og varmeovergang ved strømning av fordampende væske i horisontale rør og bend: (engl. Pressure drop and heat transfer during flow of evaporating liquid in horizontal tubes and bends): Ph.D. thesis.* Ph.D. thesis, NTH, Trondheim, Norway, Jul 1975.

P. V. Gilli. Heat transfer and pressure drop for cross flow through banks of multistart helical tubes with uniform inclinations and uniform longitudinal pitches. *Nuclear science and engineering*, 22:298–314, 1965.

H. Glaser. Wärmeübergang in Regeneratoren. *VDI Zeitschrift - Beihefte Verfahrenstechnik*, 82(4):112–125, 1938.

V. Gnielinski. Equations for calculating heat transfer in single tube rows and banks of tubes in transverse flow. *International chemical engineering*, 19(3):380–390, 1979.

V. Gnielinski. Heat transfer and pressure drop in helically coiled tubes. In *Heat transfer 1986: 1986/08/17-22, San Francisco, California*, volume 6 of *Proceedings of international heat transfer conference*, pages 2847–2854. Hemisphere Pub. Corp., 1986a. ISBN 0891165967.

G. H. Golub and C. F. VanLoan. *Matrix computations.* Johns Hopkins studies in the mathematical sciences. Johns Hopkins Univ. Press, Baltimore, Maryland, 3rd ed. edition, 1996. ISBN 0801854148.

I. D. R. Grant and D. Chisholm. Two-phase flow on the shell-side of a segmentally baffled shell-and-tube heat exchanger. *Transactions of the American Society of Mechanical Engineers*, 101:38–42, 1979.

G. Hammer, T. Lübcke, R. Kettner, M. R. Pillarella, H. Recknagel, A. Commichau, H. J. Neumann, and B. Paczynska-Lahme. Natural Gas. In *Ullmann's encyclopedia of industrial chemistry: Electronic Release 2006*. Wiley-VCH, 2006. ISBN 3527313184.

M. Hammer. Ph.D. thesis, Trondheim, Norway.

M. Hammer, S. Vist, H. Nordhus, I. L. Sperle, G. A. Owren, and O. Jørstad. Dynamic modelling of spiral wound LNG heat exchangers: Comparison to experimental results. AIChE spring national meeting, New Orleans, Louisiana, 2003/03/30-04/03.

K. M. Hangos and I. T. Cameron. The formal representation of process system modelling assumptions and their implications. *Computers and chemical engineering*, 21(S):823–828, 1997.

K. M. Hangos and I. T. Cameron. *Process modelling and model analysis*, volume 4 of *Process systems engineering*. Academic Press, San Diego, California, 2001. ISBN 0121569314.

M. Hanke, A. Olsson, K. Henrik, and M. Strömgren. Stability analysis of a reduced heat exchanger model. *Proceedings in applied mathematics and mechanics*, 5:805–806, 2005.

M. Hanke, A. Olsson, K. Henrik, and M. Strömgren. Stability analysis of a degenerate hyperbolic system modelling a heat exchanger. *Mathematics and computers in simulation*, 74:8–19, 2007.

M. M. F. Hasan, I. A. Karimi, H. E. Alfadala, and H. Grootjans. Modeling and simulation of main cryogenic heat exchanger in a base-load liquefied natural gas plant. European Symposium on Computer Aided Process Engineering, Bucharest, Romanian, 2007/05/27-30.

H. Hausen and H. Linde. *Tieftemperaturtechnik: Erzeugung sehr tiefer Temperaturen, Gasverflüssigung u. Zerlegung von Gasgemischen*. Springer, Berlin, Germany, 2nd ed. edition, 1985. ISBN 3540139729.

M. Ishii and T. Hibiki. *Thermo-fluid dynamics of two-phase flow*. Springer, New York, New York, 2006. ISBN 0387283218.

J. B. Jensen and S. Skogestad. Optimal operation of simple refrigeration cycles: Part I: Degrees of freedom and optimality of sub-cooling. *Computers and chemical engineering*, 31(5-6):712–721, 2007b.

D. Jerinic, J. Schmidt, and L. Friedel. Bewertungen von Zustandsgleichungen für Erdgase. *Chemie-Ingenieur- Technik*, 81(9):1397–1415, 2009.

D. L. V. Katz and R. L. Lee. *Natural gas engineering: Production and storage.* McGraw-Hill chemical engineering series. McGraw-Hill, New York, New York, 1990. ISBN 0070333521.

G. Kocamustafaogullari and I. Y. Chen. Falling film heat transfer analysis on a bank of horizontal tube evaporator. *American institute of chemical engineers journal*, 34(9): 1539–1549, 1988.

M. Kraume. *Transportvorgänge in der Verfahrenstechnik: Grundlagen und apparative Umsetzungen.* Engineering online library. Springer, Berlin, Germany, 2004. ISBN 3540401059.

H. Kreis. Gewickelte Wärmetauscher. In B. Thier and H. Backhaus, editors, *Apparate: Technik, Bau, Anwendung*, pages 262–264. Vulkan-Verlag, 1997. ISBN 3802721721.

A. Kröner. An engineering company's approach to filling "CAPE gaps" in process simulation. Symposium on Process Systems Engineering/European Symposium on Computer Aided Process Engineering, Garmisch-Partenkirchen, Germany, 2006/07/09-13.

T. Kronseder. *Towards nonlinear model-based online optimal control of chemical engineering plants: Parameterised controls and sensitivity functions for very large-scale index-2 DAE systems with state dependent discontinuities: Ph.D. thesis*, volume 977 of *Fortschritt-Berichte VDI Reihe 8 Meß-, Steuerungs- und Regelungstechnik.* VDI-Verlag, Düsseldorf, Germany, 2003. ISBN 3183977087.

J. M. Le Lann, A. Sargousse, P. Sere Peyrigain, and X. Joulia. Dynamic simulation of partial differential algebraic systems. ICL joint conference, Toulouse, France, 1998/07/15.

R. J. LeVeque. *Finite volume methods for hyperbolic problems*, volume 31 of *Cambridge texts in applied mathematics.* Cambridge Univ. Press, Cambridge, UK, 2002. ISBN 0521009243.

H. M. Lieberstein. *Theory of partial differential equations*, volume 93 of *Mathematics in science and engineering.* Academic Press, New York, New York, 1972. ISBN 0124495508.

K. H. Lüdtke. *Process centrifugal compressors: Basics, function, operation, design, application.* Springer, Berlin, Germany, 2004. ISBN 3540404279.

J. A. Mandler, P. A. Brochu, J. Fotopoulos, and L. Kalra. New control strategies for the LNG process. International conference & exhibition on Liquefied Natural Gas, Perth, Australia, May 1998.

W. S. Martinson. *Index and characteristic analysis partial differential equations: Ph.D. thesis.* Ph.D. thesis, Massachusetts Institute of Technology, Cambridge, Massachusetts, Feb 2000.

W. S. Martinson and P. I. Barton. A differentiation index for partial differential-algebraic equations. *SIAM journal on scientific computing*, 21(6):2295–2315, 2000.

W. S. Martinson and P. I. Barton. Distributed models in plantwide dynamic simulators. *American institute of chemical engineers journal*, 47(6):1372–1386, 2001.

W. S. Martinson and P. I. Barton. Index and characteristic analysis of linear PDAE systems. *SIAM journal on scientific computing*, 24(3):905–923, 2002.

J. M. McNaught. Two-phase forced concection heat-transfer during condensation on horizontal tube bundles. In *Heat transfer 1982: 1982/09/06-10, Munich, Gemany*, volume 5 of *Proceedings of international heat transfer conference*, pages 125–131. Hemisphere Pub. Corp., 1982.

E. Melaaen. *Dynamic simulation of the liquefaction section in baseload LNG plants: Ph.D. thesis.* Ph.D. thesis, NTH, Trondheim, Norway, Oct 1994.

K. Najim. *Process modeling and control in chemical engineering*, volume 38 of *Chemical industries*. Dekker, New York, New York, 1989. ISBN 0824782046.

B. O. Neeraas. *Condensation of hydrocarbon mixtures in coil-wound LNG heat exchangers: Tube side heat transfer and pressure drop: Ph.D. thesis.* Ph.D. thesis, NTH, Trondheim, Norway, Sep 1993.

B. O. Neeraas, A. O. Fredheim, and B. Aunan. Experimental shell-side heat transfer and pressure drop in gas flow for spiral wound LNG heat exchangers. *International journal of heat and mass transfer*, 47:3565–3572, 2004a.

B. O. Neeraas, A. O. Fredheim, and B. Aunan. Experimental data and model for heat transfer, in liquid falling film flow on shell-side, for spiral wound LNG heat exchanger. *International journal of heat and mass transfer*, 47:353–361, 2004b.

J. Neumann and C. C. Pantelides. Consistency on domain boundaries for linear PDAE systems. *SIAM journal on scientific computing*, 30(2):916–936, 2008.

C. C. Pantelides. Dynamic behaviour of process systems: Lecture notes. Feb 1998.

R. H. Perry, D. W. Grpen, and J. O. Maloney. *Perry's chemical engineers' handbook.* McGraw-Hill, New York, New York, 7th ed., int. ed. edition, 1999. ISBN 0070498415.

H. Reithmeier, M. Steinbauer, R. Stockmann, A. O. Fredheim, O. Jørstad, and P. Paurola. Industrial scale testing of a spiral wound heat exchanger. LNG conference, Doha, Qatar, 2004/04/21-24.

R. G. Sardesai, Shock, R. A. W., and D. Butterworth. Heat and mass transfer in multicomponent condensation and boiling. *Heat transfer engineering*, 3(3-4):104–114, 1982.

R. G. Sardesai, J. W. Palen, and J. Taborek. Modified resistance proration method for condensation of vapor mixtures. *AIChE symposium series*, 79:41–46, 1983.

R. K. Shah and D. P. Sekulić. *Fundamentals of heat exchanger design.* Wiley-VCH, Hoboken, New Jersey, 2003. ISBN 0471321710.

M. Shirvani and J. W. H. So. Solutions of linear differential algebraic equations. *SIAM review*, 40(2):344–346, 1998.

L. Silver. Gas cooling with aqueous condensation. *Transactions of the institution of chemical engineers*, 25:30–42, 1947.

A. Singh and M. Hovd. Mathematical modeling and control of a simple liquefied natural gas process. Scandinavian simulation society, Helsinki, Finland, 2006/09/28-29.

A. Singh and M. Hovd. Model requirement for control design of an LNG process. European Symposium on Computer Aided Process Engineering, Bucharest, Romanian, 2007/05/27-30.

M. Steinbauer and T. Hecht. Optimized calculation of helical-coiled heat exchangers in LNG plants. Eurogas, Trondheim, Norway, 1996/06/03-05.

K. Stephan and M. Abdelsalam. Heat-transfer correlations for natural convection boiling. *International journal of heat and mass transfer*, 23:73–87, 1980.

B. Thier and H. Backhaus, editors. *Apparate: Technik, Bau, Anwendung.* Vulkan-Verlag, Essen, Germany, 2nd ed edition, 1997. ISBN 3802721721.

S. Vist, M. Hammer, H. Nordhus, I. L. Sperle, G. A. Owren, and O. Jørstad. Dynamic modelling of spiral wound LNG heat exchangers: Model description. AIChE spring national meeting, New Orleans, Louisiana, 2003/03/30-04/03.

J. Voith. The production of computer-based plant optimization systems with the OPTISIM(R) process simulator: Experience with a Linde air separation plant. *Linde reports on science and technology*, 54:19–25, 1991.

G. B. Wallis. *One-dimensional two-phase flow.* McGraw-Hill, New York, New York, 1969. ISBN 0070679428.

D. Yates. Thermal efficiency: Design, lifecycle, and environmental considerations in LNG plant design. Gastech, Doha, Qatar, 2002/10/13-16.

A. Zaïm. *Dynamic optimization of an LNG plant: Case study GL2Z LNG plant in Arzew, Algeria: Ph.D. thesis*, volume 10 of *Schriftenreihe zur Aufbereitung und Veredelung.* Shaker, Aachen, Germany, Mar 2002. ISBN 3832206647.

G. Zapp and W. Sendler. Nichtlineare Steuerung und Regelung durch inverse Simulation mit komplexen DAE-Modellen. *Chemie-lngenieur- Technik*, 65(9):76, 1993.

R. Zerry. *MOSAIC, eine webbasierte Modellierungs- und Simulationsumgebung für die Verfahrenstechnik: Ph.D. thesis.* Berichte aus der Verfahrenstechnik. Shaker, Berlin, Germany, 2008. ISBN 3832271481.

3. New results on self-optimizing control

Steady state process optimization by regulation was firstly motivated by Morari *et al.* (1980). They articulated the idea that a constant setpoint policy leads to optimal operation if the underlying control structure is properly designed. Throughout this work, the term control structure refers to a set of controlled variables, *i.e.*, variables which are kept at certain setpoints by one or more controllers. The procedure of systematical development of control structures is known as control structure design (CSD). Skogestad and Postlethwaite (1996, p. 428-433) extended the idea of Morari *et al.* (1980) and gave an approximate criterion for finding (linear) control structures which inherently provide almost optimal operation. Further, they were the first which made use of the terminology self-optimizing control. Assuming a linear process model and a quadratic cost function, an exact local criterion for the identification of self-optimizing control structures was developed by Halvorsen *et al.* (2003). Both the approximate and the exact local criterion require the solution of a multivariate non-convex problem subject to structural constraints. This chapter deals with the development of new solution methods which can handle structural constraints not yet considered before.

Related work in the field of self-optimizing control is referred to in Section 3.1. Section 3.2 outlines the mathematical framework of self-optimizing control theory. In Section 3.3, a new method is proposed which is dedicated to finding self-optimizing control structures in which each controlled variable is a linear combination of the same process variable subset. As shown in Section 3.4, it can be extended to find control structures in which individual process variable subsets are mapped to each controlled variable. In Section 3.5, a numerical study is presented which aims to compare the presented CSD methods with others from the public domain. Concluding remarks are given in Section 3.6.

3.1. Related work

In this section, publicly available methods for the identification of self-optimizing control structures are reviewed. That there are a variety of CSD methods present in the open literature is due to the fact that CSD is usually a multi-objective problem. On the one hand, there is always a trade-off between structural complexity and achievable optimality. On the other hand, CSD must deal with practical aspects relating to process knowledge which cannot be made available by rules or mathematical methods.

A self-optimizing control structure subject to process variable selection can be found by exhaustive screening all possible process variable combinations. As this becomes impractical for large-scale models, Kariwala and Skogestad (2006/07/09-13); Cao and Kariwala (2008) and Kariwala and Cao (2009, 2010a) suggested the use of branch and bound algorithms to find an optimal solution. Larsson *et al.* (2001) proposed heuristical rules to decrease model size by discriminating process variables as candidate controlled variables for the sake of reducing the combinatorial diversity. They proved their approach by applying the rules on the Tennessee Eastman process, a common benchmark. Engell *et al.* (2005/07/04-08) proposed the successive disregarding of process variables as candidate controlled variables based on a maximum singular value rule applied to a characteristic matrix of the process. They applied their results to a reactive distillation column and tested it by dynamic simulation. Process variable combination methods have been published by Alstad and Skogestad (2007); Kariwala (2007); Alstad *et al.* (2009) and Kariwala *et al.* (2008) and are recalled in Appendix 3.A. These methods have in common that all controlled variables are linear combinations of the same process variable subset. The developments introduced in this chapter contribute to relax this structural constraint (Heldt, 2009, 2010a). The problem of finding the best process variable subset of certain size was proposed to be efficiently solved by either branch and bound algorithms (Kariwala and Cao, 2009, 2010a) or a mixed integer quadratic program (Yelchuru and Skogestad, 2010/07/25-28). A more efficient greedy algorithm for finding satisfactory process variable subsets was proposed by Michelsen *et al.* (2010). Apart from the linear control theory there are recent works by Jäschke *et al.* (2009/06/10-12) and Jäschke and Skogestad (2009) which focus on nonlinear self-optimizing control. As this is only applicable to processes with rather simple analytical models, it is not subject of this work.

3.2. Mathematical framework

The scheme in Figure 3.1 represents a general regulatory control structure applied to an arbitrary process plant. Based thereon the exact local criterion of Halvorsen *et al.* (2003) is introduced in this section. The vectors $\boldsymbol{u} \in \mathbb{R}^{n_u}$, $\boldsymbol{d} \in \mathbb{R}^{n_d}$, $\boldsymbol{y} \in \mathbb{R}^{n_y}$ and $\boldsymbol{c} \in \mathbb{R}^{n_u}$, respectively, correspond to the manipulated variables (MVs), disturbance variables (DVs), measured process variables (PVs) and controlled variables (CVs). A constant setpoint policy is applied. That is, the MVs are adjusted by the controller(s) until (feasibility

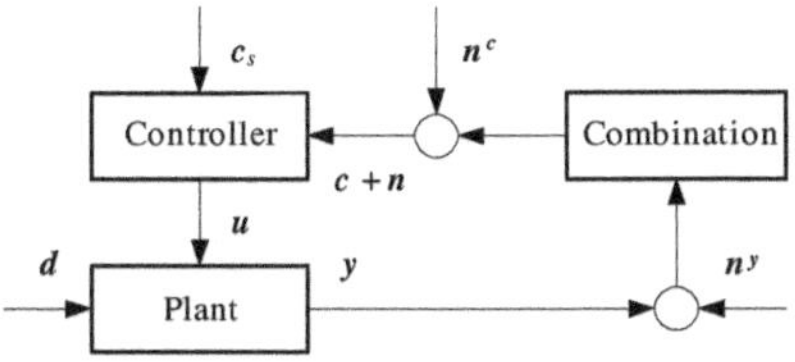

Figure 3.1.: General representation of regulatory control structures in chemical plants (after Alstad *et al.*, 2009)

provided) the CVs equal the setpoint vector $\boldsymbol{c}_{\mathrm{s}}$ whose values remain fixed. To account for measurement errors, the PVs and CVs are affected by the implementation errors $\boldsymbol{n}^{\boldsymbol{y}} \in \mathbb{R}^{n_y}$ and $\boldsymbol{n}^{\boldsymbol{c}} \in \mathbb{R}^{n_u}$.

Morari *et al.* (1980), the inventors of self-optimizing control, state that it is desirable "(...) to find a function of PVs which when held constant, leads automatically to the optimal adjustments of the MVs, and with it, the optimal operating conditions." In other words, self-optimizing control may be achieved by an appropriate mapping of PVs towards CVs, denoted by $\boldsymbol{c} = \boldsymbol{\mathfrak{H}}(\boldsymbol{y})$, where $\boldsymbol{\mathfrak{H}} \in \mathbb{R}^{n_u}$ represents the "combination" block in Figure 3.1. For deriving the exact local method, Halvorsen *et al.* (2003) considered the linear control structure $\boldsymbol{H} = \partial\boldsymbol{\mathfrak{H}}/\partial\boldsymbol{y}^T$.

The operational cost (or negative profit) of a plant denoted by J is usually affected by both types of input variables, MVs and DVs. In order to operate a plant optimally at minimum cost (or maximum profit) by compensation of DVs via direct manipulation of MVs, the problem

$$\boldsymbol{u}_{\mathrm{R/O}}(\boldsymbol{d}) = \arg\min_{\boldsymbol{u}} J(\boldsymbol{u}, \boldsymbol{d}) \;\; \text{s.t.} \;\; \boldsymbol{g}(\boldsymbol{y}, \boldsymbol{u}, \boldsymbol{d}) = 0 \tag{3.1}$$

must be solved. The solution to problem (3.1) is referred to as feed-forward re-optimization (R/O) throughout this work. Here, $\boldsymbol{g} \in \mathbb{R}^{n_y}$ denotes the steady-state model equations of the plant. For the further derivation from hereon, some simplifying assumptions need to be made.

1. The plant is in its optimal state at nominal conditions. The input/output (I/O) variables are transformed such that $\boldsymbol{u} = \boldsymbol{0}$, $\boldsymbol{d} = \boldsymbol{0}$ and $\boldsymbol{y} = \boldsymbol{0}$ hold at nominal conditions.

2. The number of MVs may be reduced as some may have no effect on the cost, others may need to be spent in "a priori" controller loops in order to either stabilize the plant or fulfill control targets such as product quality or optimally active constraints. It is assumed that $\boldsymbol{u}$ represents only the remaining MVs available for self-optimizing CSD. The "a priori" controller loops are considered part of the model equations

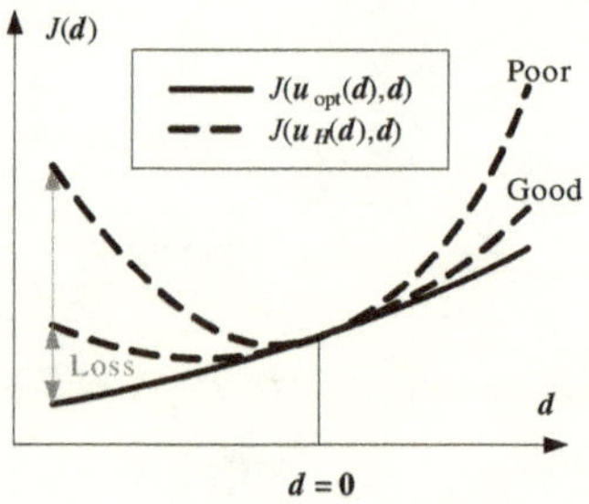

Figure 3.2.: Cost functions of regulatory control structures (dashed) and re-optimization (after Skogestad, 2000)

$\boldsymbol{g}$. Further, it is supposed that the optimally active constraints remain locally unaltered.

3. Nonlinearities of the plant are locally negligible around the optimal operating point. Consequently, the steady-state I/O model of the plant at optimal conditions can be represented by the linear equation

$$\boldsymbol{y} = \boldsymbol{G}^y \boldsymbol{u} + \boldsymbol{G}_d^y \boldsymbol{d} + \boldsymbol{n}^y, \tag{3.2}$$

where the augmented plant is given by

$$\begin{bmatrix} \boldsymbol{G}^y & \boldsymbol{G}_d^y \end{bmatrix} = -\left(\frac{\partial \boldsymbol{g}}{\partial \boldsymbol{y}^T}\right)_{\boldsymbol{0}}^{-1} \left(\frac{\partial \boldsymbol{g}}{\partial \begin{bmatrix} \boldsymbol{u}^T & \boldsymbol{d}^T \end{bmatrix}}\right)_{\boldsymbol{0}}$$

and the elements of $\boldsymbol{u}$, $\boldsymbol{d}$ and $\boldsymbol{y}$ represent deviations from nominal operating point.

4. The cost function J can be locally approximated by a second order Taylor series, *i.e.*,

$$J = J_0 + \begin{bmatrix} J_{\boldsymbol{u}} \\ J_{\boldsymbol{d}} \end{bmatrix}^T \begin{bmatrix} \boldsymbol{u} \\ \boldsymbol{d} \end{bmatrix} + \frac{1}{2} \begin{bmatrix} \boldsymbol{u} \\ \boldsymbol{d} \end{bmatrix}^T \begin{bmatrix} J_{\boldsymbol{uu}} & J_{\boldsymbol{ud}} \\ J_{\boldsymbol{ud}}^T & J_{\boldsymbol{dd}} \end{bmatrix} \begin{bmatrix} \boldsymbol{u} \\ \boldsymbol{d} \end{bmatrix}, \tag{3.3}$$

where the indices $\boldsymbol{u}$ and $\boldsymbol{d}$ indicate the partial derivative evaluated at nominal operating point. In particular, $J_{\boldsymbol{u}} = \boldsymbol{0}$ and $J_{\boldsymbol{uu}} \succ 0$ hold as a consequence of the first and second point.

Figure 3.2 shows exemplarily the operational cost of a process plant versus one DV. The cost of the feed-forward re-optimization is indicated by the solid line and gives the lower bound for feedback control with constant setpoints represented by the dashed curves. It is thus convenient to define the measure

$$L(\boldsymbol{d}) = J\left(\boldsymbol{u}_{\text{S/O}}(\boldsymbol{d}), \boldsymbol{d}\right) - J\left(\boldsymbol{u}_{\text{R/O}}(\boldsymbol{d}), \boldsymbol{d}\right) \tag{3.4}$$

known as the loss function. Here $\boldsymbol{u}_{\text{S/O}}(\boldsymbol{d})$ and $\boldsymbol{u}_{\text{R/O}}(\boldsymbol{d})$ represent the influence from DVs to MVs for feedback control with constant setpoints and re-optimization, respectively. A good self-optimizing (S/O) control structure refers to a situation in which the functions $\boldsymbol{u}_{\text{S/O}}(\boldsymbol{d})$ and $\boldsymbol{u}_{\text{R/O}}(\boldsymbol{d})$ are similar to each other and the loss $L(\boldsymbol{d})$ is small for the expected disturbances.

The dependency of (3.4) from the feedback control structure is not yet apparent and will be derived below. For shorter notation, the indication of the dependency on $\boldsymbol{d}$ is dropped in the sequel. Insertion of (3.3) into (3.4) yields

$$L = \frac{1}{2}\boldsymbol{u}_{\text{S/O}}^T J_{\boldsymbol{uu}}\, \boldsymbol{u}_{\text{S/O}} + \boldsymbol{u}_{\text{S/O}} J_{\boldsymbol{ud}}\, \boldsymbol{d} - \boldsymbol{u}_{\text{R/O}} J_{\boldsymbol{ud}}\, \boldsymbol{d} - \frac{1}{2}\boldsymbol{u}_{\text{R/O}}^T J_{\boldsymbol{uu}}\, \boldsymbol{u}_{\text{R/O}}, \tag{3.5}$$

for which $J_{\boldsymbol{u}} = \boldsymbol{0}$ was taken into account. Using the first order Taylor series

$$\frac{\partial J}{\partial \boldsymbol{u}} = \underbrace{J_{\boldsymbol{u}}}_{=\boldsymbol{0}} + J_{\boldsymbol{uu}}\, \boldsymbol{u} + J_{\boldsymbol{ud}}\, \boldsymbol{d}$$

and the fact that re-optimization implies $\partial J / \partial \boldsymbol{u} = 0$, it follows that

$$\boldsymbol{u}_{\text{R/O}} = -J_{\boldsymbol{uu}}^{-1} J_{\boldsymbol{ud}}\, \boldsymbol{d}. \tag{3.6}$$

Using this result, (3.5) can be transformed into

$$L = \frac{1}{2}\boldsymbol{z}^T \boldsymbol{z} \tag{3.7}$$

with the loss variables

$$\boldsymbol{z} = J_{\boldsymbol{uu}}^{1/2}\left(\boldsymbol{u}_{\text{S/O}} - \boldsymbol{u}_{\text{R/O}}\right). \tag{3.8}$$

In (3.8), $J_{\boldsymbol{uu}}^{1/2}$ is the Cholesky factor corresponding to $J_{\boldsymbol{uu}} = \left(J_{\boldsymbol{uu}}^{1/2}\right)^T J_{\boldsymbol{uu}}^{1/2}$. From $\boldsymbol{H}\,\boldsymbol{y} = \boldsymbol{c} \overset{!}{=} \boldsymbol{c}_{\text{s}} = \boldsymbol{0}$ and (3.2) it can be obtained that

$$\boldsymbol{u}_{\text{S/O}} = -\left(\boldsymbol{H}\,\boldsymbol{G}^{y}\right)^{-1} \boldsymbol{H}\left(\boldsymbol{G}_d^{y}\,\boldsymbol{d} + \boldsymbol{n}^{y}\right). \tag{3.9}$$

Insertion of (3.9) and (3.6) into (3.8) yields

$$\boldsymbol{z} = -J_{\boldsymbol{uu}}^{1/2}\left(\boldsymbol{H}\,\boldsymbol{G}^{y}\right)^{-1} \boldsymbol{H}\left(\left(\boldsymbol{G}_d^{y} - \boldsymbol{G}^{y} J_{\boldsymbol{uu}}^{-1} J_{\boldsymbol{ud}}\right)\boldsymbol{d} + \boldsymbol{n}^{y}\right).$$

For simplification of notation, the matrices

$$\boldsymbol{G}_{\boldsymbol{z}}^{\boldsymbol{y}} = \boldsymbol{G}^{\boldsymbol{y}} \boldsymbol{J}_{\boldsymbol{uu}}^{-1/2} \tag{3.10a}$$

$$\tilde{\boldsymbol{F}} = \left[\begin{array}{cc} \left(\boldsymbol{G}_{\boldsymbol{d}}^{\boldsymbol{y}} - \boldsymbol{G}^{\boldsymbol{y}} \boldsymbol{J}_{\boldsymbol{uu}}^{-1} \boldsymbol{J}_{\boldsymbol{ud}}\right) \boldsymbol{W}_{\boldsymbol{d}} & \boldsymbol{W}_{\boldsymbol{n}^{\boldsymbol{y}}} \end{array} \right] \tag{3.10b}$$

$$\boldsymbol{M} = (\boldsymbol{H}\, \boldsymbol{G}_{\boldsymbol{z}}^{\boldsymbol{y}})^{-1}\, \boldsymbol{H}\, \tilde{\boldsymbol{F}} \tag{3.10c}$$

are introduced which yields the short expression

$$\boldsymbol{z} = \boldsymbol{M}\, \boldsymbol{f}. \tag{3.11}$$

Here,

$$\boldsymbol{f} \in \mathcal{F} = \left\{ \boldsymbol{x} \in \mathbb{R}^{n_d + n_y} : \|\boldsymbol{x}\|_2 \leq 1 \right\}$$

combines the scaled disturbances and implementation errors and $\boldsymbol{W}_{\boldsymbol{d}}$ and $\boldsymbol{W}_{\boldsymbol{n}^{\boldsymbol{y}}}$ are the respective diagonal scaling matrices.

From (3.7) and (3.11), Halvorsen *et al.* (2003) concluded that the worst-case loss for feedback control is given by

$$L_{\mathrm{wc}} = \max_{\boldsymbol{f} \in \mathcal{F}} L = \frac{1}{2}\, \bar{\sigma}^2 (\boldsymbol{M}) \tag{3.12}$$

where $\bar{\sigma}$ indicates the largest singular value. Consequently, a self-optimizing control structure with least worst-case loss may be obtained by solving the problem

$$\boldsymbol{H} = \arg \min_{\boldsymbol{H}} \bar{\sigma} (\boldsymbol{M}). \tag{3.13}$$

An upper bound for $\bar{\sigma}(\boldsymbol{M})$ can be derived, *i.e.*,

$$\bar{\sigma} \left(\boldsymbol{H}\, \tilde{\boldsymbol{F}}\right) / \underline{\sigma} (\boldsymbol{H}\, \boldsymbol{G}_{\boldsymbol{z}}^{\boldsymbol{y}}) \geq \bar{\sigma} (\boldsymbol{M}),$$

where $\underline{\sigma}$ indicates the smallest singular value. Accordingly, approximate solutions can be found by either minimizing $\bar{\sigma}\left(\boldsymbol{H}\, \tilde{\boldsymbol{F}}\right)$ or maximizing $\underline{\sigma}(\boldsymbol{H}\, \boldsymbol{G}_{\boldsymbol{z}}^{\boldsymbol{y}})$. The first strategy is made use of by the nullspace method (Alstad and Skogestad, 2007), while the second strategy gives rise to the maximum singular value (MSV) rule, which reads

$$\boldsymbol{H} = \arg \max_{\boldsymbol{H}} \underline{\sigma} (\boldsymbol{H}\, \boldsymbol{G}_{\boldsymbol{z}}^{\boldsymbol{y}}) \tag{3.14}$$

when scaling is disregarded. Both strategies have the advantage that they reduce complexity and are thus favored by several authors. Kariwala *et al.* (2008) proposed the average loss

$$L_{\text{av}} = \int_{\boldsymbol{f} \in \mathcal{F}} L \, \mathrm{d}\boldsymbol{f} = \frac{1}{6\left(n_{\mathcal{Y}_{\text{u}}} + n_{\boldsymbol{d}}\right)} \|\boldsymbol{M}\|_{\text{F}}^2 \tag{3.15}$$

based on the assumption that $\|\boldsymbol{f}\|_2$ is uniformly distributed over the range $0 \leq \|\boldsymbol{f}\|_2 \leq 1$. Here, $\|.\|_{\text{F}}$ indicates the Frobenius norm and $n_{\mathcal{Y}_{\text{u}}} = |\mathcal{Y}_{\text{u}}|$ where $\mathcal{Y}_{\text{u}} = \mathcal{Y}_1 \cap \cdots \cap \mathcal{Y}_{n_u}$ and $\mathcal{Y}_i$ is the PV subset corresponding to the ith CV. *I.e.*, $\mathcal{Y}_{\text{u}}$ indicates the overall subset of all PVs who have at least one nonzero linear coefficient in the control structure (*e.g.*, $n_{\mathcal{Y}_{\text{u}}} = n_{\boldsymbol{u}}$ for PV selection). Kariwala *et al.* (2008) state that the average loss (3.15) is usually a better estimate of the loss as the worst-case loss (3.12) tends to overestimation. According to (3.15), Kariwala *et al.* (2008) suggested solving

$$\boldsymbol{H} = \arg\min_{\boldsymbol{H}} \|\boldsymbol{M}\|_{\text{F}} \tag{3.16}$$

instead of (3.13). They proved that the solution to (3.16) is super-optimal in the sense that it minimizes both the average and the worst-case loss.

Remark 3.1. According to (3.15), the average loss depends on the number of PVs used as combination variables $n_{\mathcal{Y}_{\text{u}}}$. This dependency was omitted in (3.16) in order to limit the favor of large $n_{\mathcal{Y}_{\text{u}}}$.

After this brief survey it can be concluded that the identification of self-optimizing control structures can be achieved by solving (3.13), (3.14) or (3.16) subject to certain structural constraints. An optimal control structure subject to PV selection can be found by exhaustive screening of all possible PV combinations. As this becomes impractical for large-scale models, several authors (Kariwala and Skogestad, 2006/07/09-13; Cao and Kariwala, 2008; Kariwala and Cao, 2009, 2010a) suggested the use of efficient branch and bound (BAB) algorithms. Others (Alstad and Skogestad, 2007; Kariwala, 2007; Kariwala *et al.*, 2008; Alstad *et al.*, 2009)[1] published methods for the identification of structures in which each CV is a linear combination of an a priori imposed PV subset (recalled in Appendix 3.A). In order to get the best structure subject to a certain PV set size, one possibility is the exhaustive search, *i.e.*, screening of all possible PV subsets and for each subset applying one of these methods. As for the PV selection problem, this becomes computationally expensive for large $n_{\boldsymbol{y}}$ and also for large PV subset sizes smaller $n_{\boldsymbol{y}}/2$. Therefore, Kariwala and Cao (2009, 2010a) proposed efficient BAB solution methods for finding the best PV subset subject to a certain set size.

This thesis is dedicated to the solution of (3.13), (3.14) or (3.16) subject to more advanced structural constraints than PV selection or linear combinations of a common

[1] Alstad *et al.* (2009) proposed two different methods, the extended nullspace method and one which was not named but for which a linearly constrained quadratic problem needed to be solved (see Appendix 3.B). If not explicitly stated, the method by Alstad *et al.* (2009) refers to the latter method throughout this work.

PV subset with certain set size. This is the concern of Section 3.4. However, these methods are based on a further method for common PV subset combination named GSVD method. It was priorly presented by Heldt (2009, 2010a) and is recalled in Section 3.3.

3.3. The GSVD method

Suppose problems (3.13) and (3.16) need to be solved subject to a predefined structure in which a linear combination of the same PV subset $\mathcal{Y}$ with set size $n_{\mathcal{Y}} = |\mathcal{Y}| \geq n_{\boldsymbol{u}}$ is mapped towards every CV. First, it is convenient to restate (3.11) as

$$\boldsymbol{z}^T \left(\boldsymbol{G}_{\boldsymbol{z}}^{\mathcal{Y}}\right)^T \boldsymbol{H}_{\mathcal{Y}}^T = \boldsymbol{f}^T \left(\tilde{\boldsymbol{F}}^{\mathcal{Y}}\right)^T \boldsymbol{H}_{\mathcal{Y}}^T. \tag{3.17}$$

Here, the sub and superscript $\mathcal{Y}$ denotes that relating columns in $\boldsymbol{H}$ and rows in $\boldsymbol{G}_{\boldsymbol{z}}^{\boldsymbol{y}}$ and $\tilde{\boldsymbol{F}}$ are extracted. According to Paige and Saunders (1981), a generalized singular value decomposition[2] (GSVD) of the matrix pair

$$\left\{\left(\tilde{\boldsymbol{F}}^{\mathcal{Y}}\right)^T, \left(\boldsymbol{G}_{\boldsymbol{z}}^{\mathcal{Y}}\right)^T\right\}$$

exists which is used to rewrite (3.17) as

$$\boldsymbol{z}^T \boldsymbol{U} \boldsymbol{\Sigma} \boldsymbol{V}^T \boldsymbol{H}_{\mathcal{Y}}^T = \boldsymbol{f}^T \tilde{\boldsymbol{U}} \tilde{\boldsymbol{\Sigma}} \boldsymbol{V}^T \boldsymbol{H}_{\mathcal{Y}}^T. \tag{3.18}$$

The properties of the decomposed matrices are pointed out next. The matrices $\boldsymbol{U} \in \mathbb{R}^{n_{\boldsymbol{u}} \times n_{\boldsymbol{u}}}$ and $\tilde{\boldsymbol{U}} \in \mathbb{R}^{n_{\boldsymbol{f}} \times n_{\boldsymbol{f}}}$ are unitary, *i.e.*, $\boldsymbol{U}^T \boldsymbol{U} = \boldsymbol{I}_{n_u}$ and $\tilde{\boldsymbol{U}}^T \tilde{\boldsymbol{U}} = \boldsymbol{I}_{n_f}$. The matrix $\boldsymbol{V} \in \mathbb{R}^{n_{\mathcal{Y}} \times q}$ owns the property that it has full column rank, *i.e.*, $\operatorname{rank}(\boldsymbol{V}) = q$, whereas $q = \operatorname{rank}\left(\left[\begin{array}{cc} \tilde{\boldsymbol{F}}^{\mathcal{Y}} & \boldsymbol{G}_{\boldsymbol{z}}^{\mathcal{Y}} \end{array}\right]\right)$. The matrix $\boldsymbol{\Sigma} \in \mathbb{R}^{n_{\boldsymbol{u}} \times q}$ is trailing diagonal, with $\boldsymbol{\Sigma}^T \boldsymbol{\Sigma} = \operatorname{diag}\left(\beta_1^2, \beta_2^2, ..., \beta_q^2\right)$ and $0 \leq \beta_i \leq \beta_{i+1} \leq 1$. The matrix $\tilde{\boldsymbol{\Sigma}} \in \mathbb{R}^{n_{\boldsymbol{f}} \times q}$ is leading diagonal, with $\tilde{\boldsymbol{\Sigma}}^T \tilde{\boldsymbol{\Sigma}} = \operatorname{diag}\left(\alpha_1^2, \alpha_2^2, ..., \alpha_q^2\right)$ and $1 \geq \alpha_i \geq \alpha_{i+1} \geq 0$. The relation $\alpha_i^2 + \beta_i^2 = 1 \forall i \in \{1, \ldots, q\}$ holds. Further, the number of $r = q - \operatorname{rank}\left(\boldsymbol{G}_{\boldsymbol{z}}^{\mathcal{Y}}\right)$ leading α_i and β_i are 1 and 0, respectively, and the number of $s = q - \operatorname{rank}\left(\tilde{\boldsymbol{F}}^{\mathcal{Y}}\right)$ trailing α_i and β_i are 0 and 1, respectively. The quotient $\sigma_i = \alpha_i / \beta_i$ is referred to as the ith largest generalized singular value of the matrix pair $\left\{\left(\tilde{\boldsymbol{F}}^{\mathcal{Y}}\right)^T, (\boldsymbol{G}_{\boldsymbol{z}}^{\boldsymbol{y}})_{\mathcal{Y}}^T\right\}$. For more information on GSVD and on how the resulting matrices can be computed, the reader is referred to standard linear algebra textbooks, *e.g.*, the textbooks of Hogben (2007, pp. 75.20-24) and Golub and VanLoan (1996, pp. 465-467) and the MATLAB® documentation (2008b) of gsvd().

[2] There are different types of GSVDs. Some require a rank condition on the matrix pair, others not. Here the latter type is referred to, corresponding to the MATLAB® function gsvd().

Lemma 3.2. *Given two arbitrary matrices $\boldsymbol{A} \in \mathbb{R}^{n\times p}$ with $n \geq p$ and $\boldsymbol{B} \in \mathbb{R}^{m\times p}$, it holds that*

$$\left\| \begin{bmatrix} \boldsymbol{A} \\ \boldsymbol{B} \end{bmatrix} \right\|_\star \geq \|\boldsymbol{A}\|_\star$$

for $\star$ indicating either induced two-norm or Frobenius norm.

Proof. The proof follows trivially from recursive application of Horn and Johnson (1990, Theorem 7.3.9). □

Theorem 3.3. *If* $\operatorname{rank}\left(\boldsymbol{G}_{\boldsymbol{z}}^{\mathcal{Y}}\right) = n_{\boldsymbol{u}}$, *then the minimum worst-case and average loss are given by*

$$L_{wc} = \frac{1}{2}\sigma_{r+1}^2 \tag{3.19}$$

and

$$L_{av} = \frac{1}{6\,(n_{\mathcal{Y}} + n_{\boldsymbol{d}})} \sum_{i=r+1}^{q} \sigma_i^2, \tag{3.20}$$

respectively. (3.19) *and* (3.20) *are obtained by selecting*

$$\boldsymbol{H}_{\mathcal{Y}} = \boldsymbol{M}_n\, \boldsymbol{P}_2^T,$$

where $\boldsymbol{M}_n \in \mathbb{R}^{n_{\boldsymbol{u}}\times n_{\boldsymbol{u}}}$ is an arbitrary regular matrix and $\boldsymbol{P}_2$ is the $n_{\mathcal{Y}} \times n_{\boldsymbol{u}}$ sub-block of $\boldsymbol{P} = \begin{bmatrix} \boldsymbol{P}_1 & \boldsymbol{P}_2 \end{bmatrix} = \left(\boldsymbol{V}^T\right)^\dagger = \boldsymbol{V}\left(\boldsymbol{V}^T\boldsymbol{V}\right)^{-1}$ in which $\dagger$ denotes the right pseudo-inverse (can be replaced by the inverse if $q = n_{\mathcal{Y}}$ holds).

Proof. It is convenient to rewrite (3.18) as

$$\boldsymbol{z}^T\boldsymbol{U}\begin{bmatrix} \boldsymbol{0}_{n_{\boldsymbol{u}}\times r} & \boldsymbol{\Sigma}_2 \end{bmatrix}\begin{bmatrix} \boldsymbol{V}_1^T \\ \boldsymbol{V}_2^T \end{bmatrix}\boldsymbol{H}_{\mathcal{Y}}^T = \boldsymbol{f}^T\tilde{\boldsymbol{U}}\begin{bmatrix} \tilde{\boldsymbol{\Sigma}}_1 & \boldsymbol{0}_{r\times n_{\boldsymbol{u}}} \\ \boldsymbol{0}_{n_{\boldsymbol{u}}\times r} & \tilde{\boldsymbol{\Sigma}}_2 \\ \boldsymbol{0}_{\bar{s}\times r} & \boldsymbol{0}_{\bar{s}\times n_{\boldsymbol{u}}} \end{bmatrix}\begin{bmatrix} \boldsymbol{V}_1^T \\ \boldsymbol{V}_2^T \end{bmatrix}\boldsymbol{H}_{\mathcal{Y}}^T, \tag{3.21}$$

where $\bar{s} = \max\left(0, n_{\boldsymbol{f}} - q\right)$ and $\boldsymbol{V}$ is split into two sub-blocks $\boldsymbol{V}_1 \in \mathbb{R}^{n_{\mathcal{Y}}\times r}$ and $\boldsymbol{V}_1 \in \mathbb{R}^{n_{\mathcal{Y}}\times n_{\boldsymbol{u}}}$ such that

$$\boldsymbol{V}^T\boldsymbol{P} = \begin{bmatrix} \boldsymbol{V}_1^T\boldsymbol{P}_1 & \boldsymbol{V}_1^T\boldsymbol{P}_2 \\ \boldsymbol{V}_2^T\boldsymbol{P}_1 & \boldsymbol{V}_2^T\boldsymbol{P}_2 \end{bmatrix} = \begin{bmatrix} \boldsymbol{I}_{r\times r} & \\ & \boldsymbol{I}_{n_{\boldsymbol{u}}\times n_{\boldsymbol{u}}} \end{bmatrix}. \tag{3.22}$$

The condition $\operatorname{rank}\left(\boldsymbol{G}_{\boldsymbol{z}}^{\mathcal{Y}}\right) = n_{\boldsymbol{u}}$ implies that $\boldsymbol{\Sigma}_2$ is regular. Thus, a finite $\|\boldsymbol{M}\|_\star$ imposes the necessary condition that $\boldsymbol{V}_2^T\boldsymbol{H}_{\mathcal{Y}}^T$ is regular. If this is provided, it follows from (3.21) that

$$\|\boldsymbol{M}\|_\star = \left\| \begin{bmatrix} \tilde{\boldsymbol{\Sigma}}_1\boldsymbol{V}_1^T\boldsymbol{H}_{\mathcal{Y}}^T\left(\boldsymbol{V}_2^T\boldsymbol{H}_{\mathcal{Y}}^T\right)^{-1}\boldsymbol{\Sigma}_2^{-1} \\ \tilde{\boldsymbol{\Sigma}}_2\boldsymbol{\Sigma}_2^{-1} \end{bmatrix} \right\|_\star.$$

Here, it was used that induced two-norm and Frobenius norm are both unitarily invariant. From Lemma 3.2 it can be concluded that a $\boldsymbol{H}_{\mathcal{Y}}^T$ independent lower bound on $\|\boldsymbol{M}\|_\star$ exists, *i.e.*,

$$\|\boldsymbol{M}\|_\star \geq \left\|\tilde{\boldsymbol{\Sigma}}_2 \boldsymbol{\Sigma}_2^{-1}\right\|_\star .$$

As the condition $\boldsymbol{V}_1^T \boldsymbol{H}_{\mathcal{Y}}^T = \boldsymbol{0}$ meets this lower bound, it is sufficient for minimizing $\|\boldsymbol{M}\|_\star$ and the results (3.12) and (3.15) can be trivially derived. As can be seen from (3.22), the selection $\boldsymbol{H}_{\mathcal{Y}} = \boldsymbol{M}_n \boldsymbol{P}_2^T$ satisfies both the necessary condition that $\boldsymbol{V}_2^T \boldsymbol{H}_{\mathcal{Y}}^T$ is regular and the sufficient condition $\boldsymbol{V}_1^T \boldsymbol{H}_{\mathcal{Y}}^T = \boldsymbol{0}$. □

Remark 3.4. For perfect disturbance rejection, *i.e.*, $\boldsymbol{M} = \boldsymbol{0}$, it is required that at least $n_{\boldsymbol{u}}$ trailing α_i are zero. As indicated above, the number of $s = q - \operatorname{rank}\left(\tilde{\boldsymbol{F}}^{\mathcal{Y}}\right)$ trailing α_i are in fact zero. Thus, perfect disturbance rejection occurs if the inequality $s \geq n_{\boldsymbol{u}}$ is satisfied. Due to $q \leq n_{\mathcal{Y}}$, the sufficient condition for perfect disturbance rejection reads

$$n_{\mathcal{Y}} \geq n_{\boldsymbol{u}} + \operatorname{rank}\left(\tilde{\boldsymbol{F}}^{\mathcal{Y}}\right).$$

The necessary condition is thus $n_{\mathcal{Y}} > \operatorname{rank}\left(\tilde{\boldsymbol{F}}^{\mathcal{Y}}\right)$, which can only be satisfied if a certain number of implementation errors are zero. If all implementation errors are disregarded and each DV has a linearly independent effect on the PVs of the set $\mathcal{Y}$, then $\operatorname{rank}\left(\tilde{\boldsymbol{F}}^{\mathcal{Y}}\right) = n_{\boldsymbol{d}}$ and the condition for perfect disturbance rejection reads $n_{\mathcal{Y}} \geq n_{\boldsymbol{u}} + n_{\boldsymbol{d}}$. This is in agreement with results of other combination methods (Alstad and Skogestad, 2007; Kariwala, 2007; Kariwala *et al.*, 2008; Alstad *et al.*, 2009).

The equivalence of the GSVD method with previously published methods is pointed out in Appendix 3.B.

3.4. Advanced control structures

This section is dedicated to the solution of (3.13), (3.14) or (3.16) subject to advanced structural constraints. Principle ideas convenient for the subsequent discussion are presented in Section 3.4.1. In Section 3.4.2, it is argued why the common PV subsets are considered insufficient for application in industrial practice. Section 3.4.3 deals with the major obstacles faced when taking advanced structures into account. The concern of Section 3.4.4 is an appropriate transformation of the solution space for problems (3.13) and (3.16) subject to an arbitrary predefined linear control structure. Approximate solution approaches which make use of this transformation are presented in Section 3.4.5.

3.4.1. Principles

For clearer expressiveness, it is convenient to provide definitions of control structure types and sets of control structures.

Definition 3.5. A control structure in which all CVs are linear combinations of the same PV subset $\mathcal{Y} \subseteq \mathcal{Y}_{\boldsymbol{y}}$ of size $n_{\mathcal{Y}} = |\mathcal{Y}|$ is referred to as a *column control structure*. A *common-sized control structure* refers to a control structure in which the ith CV is a linear combination of an individual PV subset $\mathcal{Y}_i$ with the constraint that all PV subsets have the same set size $n_{\text{cs}} = |\mathcal{Y}_i| \, \forall i \in \{1, \ldots, n_{\boldsymbol{u}}\}$. In a more general *individually-sized control structure*, the ith CV is a linear combination of an individual PV subset $\mathcal{Y}_i$ with individual set size $n_{\mathcal{Y}_i} = |\mathcal{Y}_i|$.

Definition 3.6. The largest set of structural solutions which fulfills a set of constraints is referred to as *superstructure*. One structural solution within the superstructure is called an *element* of the superstructure.

Some properties of column control structures are presented next.

Lemma 3.7. *Let* $\boldsymbol{H}$ *represent a column control structure of size* $n_{\mathcal{Y}}$ *with finite worst-case/average loss,* i.e., *the necessary condition* $\operatorname{rank}(\boldsymbol{H}) = n_{\boldsymbol{u}}$ *holds. Then, for every* $\boldsymbol{H}$ *there exists a common-sized control structure* $\boldsymbol{H}_{cs}$ *with a PV subset size per CV of* $n_{cs} = n_{\mathcal{Y}} - (n_{\boldsymbol{u}} - 1)$ *and the same worst-case/average loss as* $\boldsymbol{H}$.

Proof. Without loss of generality, it is assumed that the PVs are reordered such that $\boldsymbol{H} = \begin{bmatrix} \boldsymbol{H}_1 & \boldsymbol{H}_2 & \boldsymbol{0}_{n_{\boldsymbol{u}} \times (n_{\boldsymbol{y}} - n_{\mathcal{Y}})} \end{bmatrix}$ where $\boldsymbol{H}_1 \in \mathbb{R}^{n_{\boldsymbol{u}} \times n_{\boldsymbol{u}}}$ is regular and $\boldsymbol{H}_2 \in \mathbb{R}^{n_{\boldsymbol{u}} \times (n_{\mathcal{Y}} - n_{\boldsymbol{u}})}$. Now, the fact is used that $\boldsymbol{M}(\boldsymbol{H}) = \boldsymbol{M}(\boldsymbol{M}_n \boldsymbol{H})$, where $\boldsymbol{M}_n \in \mathbb{R}^{n_{\boldsymbol{u}} \times n_{\boldsymbol{u}}}$ may be any regular matrix. This is self-evident by observation of (3.10c). Accordingly, $\boldsymbol{H}$ and

$$\boldsymbol{H}_{\text{cs}} = \boldsymbol{H}_1^{-1} \boldsymbol{H} = \begin{bmatrix} \boldsymbol{I}_{n_{\boldsymbol{u}}} & \boldsymbol{H}_1^{-1} \boldsymbol{H}_2 & \boldsymbol{0}_{n_{\boldsymbol{u}} \times (n_{\boldsymbol{y}} - n_{\mathcal{Y}})} \end{bmatrix}$$

have equal worst-case/average loss (3.12)/(3.15). It is obvious that $\boldsymbol{H}_{\text{cs}}$ has $n_{\text{cs}} = n_{\mathcal{Y}} - (n_{\boldsymbol{u}} - 1)$ nonzero elements per row. □

Corollary 3.8. *Let* $\boldsymbol{H}$ *be a column control structure with PV subset size* $n_{\mathcal{Y}} = n_{\boldsymbol{u}}$ *and finite loss. Then, the worst-case/average loss is independent of the coefficients in* $\boldsymbol{H}$. *Rather, the worst-case/average loss depends only on the selection of the PV subset* $\mathcal{Y}$.

Proof. According to Lemma 3.7, it exists a common-sized control structure $\boldsymbol{H}_{\text{cs}}$ with the same worst-case/average loss as $\boldsymbol{H}$ and a PV subset size of $n_{\text{cs}} = 1$. □

Remark 3.9. Lemma 3.7 states that common-sized control structures can be calculated from (optimal) column control structures by simple linear transformation. It is important to stress that these results are not necessarily optimal in the sense to problems (3.13), (3.14) and (3.16) subject to common-sized superstructures.

Remark 3.10. As indicated in Lemma 3.7, the worst-case/average loss of a control structure $\boldsymbol{H}$ is invariant with respect to linear transformations of the form $\boldsymbol{M}_n \boldsymbol{H}$, where $\boldsymbol{M}_n \in \mathbb{R}^{n_u \times n_u}$ may be any regular matrix. However, the structure of $\boldsymbol{H}$ is not generally invariant with respect to such transformations, only if $\boldsymbol{M}_n$ is a diagonal matrix.

3.4.2. Motivation

Some advantages of common-/individually-sized control structures over column control structures are pointed out below.

1. A smaller PV subset size per CV is on the one hand favorable due to better practical acceptance, but on the other hand usually accompanied by a larger worst-case/average loss. From Lemma 3.7 it can be concluded that for $n_{\boldsymbol{u}} > 1$ a reduction in PV subset size per CV without affecting the worst-case/average loss can be achieved if, instead of a column control structure, a common-sized control structure is taken into account. Accordingly, a common-sized control structure needs less degrees of freedom (number of structural nonzeros) in $\boldsymbol{H}$ than a column control structure to achieve the same loss.

2. By implication of the first argument, it is evident that a smaller worst-case/average loss can be achieved if, instead of a column control structure, a common-sized control structure with equal PV subset size per CV is taken into account. The numerical demonstration of this fact is given in Section 3.5.

3. By definition, $n_{\mathcal{Y}_\mathrm{u}} = n_{\mathcal{Y}}$ for column control structures and $n_{\mathcal{Y}_\mathrm{u}} > n_\mathrm{cs}$ for common-sized control structures. Thus, the size of the union PV subset $n_{\mathcal{Y}_\mathrm{u}}$ of a common-sized control structure is larger than that of a column control structure if both have the same PV subset size per CV, *i.e.*, if $n_{\mathcal{Y}} = n_\mathrm{cs}$. This has two implications. First, a larger $n_{\mathcal{Y}_\mathrm{u}}$ leads to an increased chance that noise cancels out which makes the control structure more robust in terms measurement failure/loss. Second, a larger $n_{\mathcal{Y}_\mathrm{u}}$ favors smaller average loss. This is due to the multiplier $(n_{\mathcal{Y}_\mathrm{u}} + n_{\boldsymbol{d}})^{-1}$ in expression (3.15) and amplifies the second argument regarding average loss.

4. For $n_{\mathcal{Y}} = n_{\boldsymbol{u}}$ the optimality of column control structures is only a matter of PV subset selection as pointed out in Corollary 3.8. Consequently, the degrees of freedom of the linear coefficients are in this case unused and a PV selection structure, *i.e.*, a common-sized control structures with $n_\mathrm{cs} = 1$, is favorable.

5. Column control structures $\boldsymbol{H}$ generally fail if $n_{\boldsymbol{y}} < n_{\boldsymbol{u}}$ holds. This is due to the fact that rank$(\boldsymbol{H}) < n_{\boldsymbol{u}}$ which leads to a singular $\boldsymbol{H}\,\boldsymbol{G}_{\boldsymbol{z}}^{\boldsymbol{y}}$ and, by observation of (3.11) and (3.10c), to infinite loss. In contrast, common-sized and individually-sized control structures do not necessarily fail if $n_{\mathrm{cs}} < n_{\boldsymbol{u}}$ and $n_{\boldsymbol{y}_i} < n_{\boldsymbol{u}}$ for all/any $i \in \{1, \ldots, n_{\boldsymbol{u}}\}$, respectively. Assume for this argument that $n_{\boldsymbol{u}} > 1$.

6. Some aspects of practical acceptance require taking individual PV subset per CV into account as in common-/individually-sized control structures.

   a) I/O selection based on heuristic rules is a common practice. If decentralized controllers are used, physical closeness between CVs and MVs is probably the most common rule, in order to achieve good cause and effect and short delays between MVs and CVs. If the MVs are far apart from each other (*e.g.*, in large scale processes), then physical closeness can only be achieved by taking individual PV subsets per CV into account.

   b) Mixed-unit PV combinations, *i.e.*, combinations including different physical units, are not recommended to be utilized due to poor practical acceptance. If pure-unit combinations are imposed, then each CV must be of the same unit when column control structures are considered and thus all information of PVs outside the selected group must be disregarded. In common/individually-sized control structures more information can be taken into account as the unit of each CV can be different.

   c) Usually, commissioning of a whole plant occurs by taking sub-units subsequently into operation according to a well defined sequence. It is thus necessary that the CVs of each sub-unit consist only of PVs which relate to the same sub-unit or sub-units already in operation. In this respect is also beneficial to have different PV subsets per CV.

From these points it can be concluded that it is very desirable to solve the problems (3.13), (3.14) and (3.16) subject to common/individually-sized control structures. That this is a rather challenging task is pointed out in the subsequent section.

3.4.3. Obstacles

The difficulty of solving problems (3.13), (3.14) and (3.16) depends on the imposed structural constraint. For PV selection, the problems turn into pure integer problems. If a single predefined column structure is imposed, then the average loss problem (3.16) can be transformed into a linearly convex quadratic problem (see Alstad *et al.*, 2009) for which an analytical solution can be derived which also solve the worst-case loss problem (3.13).

Based on this result, it was shown by Kariwala and Cao (2009, 2010a) that problems (3.13) and (3.16) subject to column superstructures with certain PV subset sizes turn into pure integer problems.

Considering common/individually-sized structures, different constraint types can be imposed which then lead to different solution techniques. One can either impose a predefined structure or a superstructure with certain size(s), n_{cs} and n_i, of the PV subsets. In the latter case auxiliary conditions could be required by the user. For instance, only certain PVs may be allowed to be combined with each other (*e.g.*, according to their physical units). Further, an individual candidate PV set may be imposed for each CV (*e.g.*, from different plant locations). A weaker condition would be to prohibit overlapping of the PV subsets.

Remark 3.11. It is important to note that the development of a general framework which covers all these constraints is an issue in itself and is not within the scope of this thesis. For simplicity, only common/individually-sized control superstructures with the same candidate PV subsets for every CV are taken into account here.

A general explicit solution to one of the problems (3.13), (3.14) and (3.16) subject to an arbitrary predefined common/individually-sized structure could not be derived by the author of this thesis and is conjectured that it is non-existent. Consequently, solutions can only be obtained by applying iterative solution techniques. However, this is cumbersome due to non-convexity. The iterative solution of problem (3.13) is even more complex as it states a min/max problem.

Due to the lack of an explicit solution for problems (3.13), (3.14) and (3.16) subject to a predefined common/individually-sized structure, the same problems subject to common/individually-sized superstructures state mixed-integer problems. *I.e.*, for each element of the superstructure an iterative solution approach must be applied. This becomes impractical for large number of elements in the superstructure due to computational overload. The following example shows that the number of elements in common/individually-sized control structures can be huge even for small system dimensions.

Example 3.12. Suppose that a common-sized superstructure with PV set size n_{cs} is imposed as a constraint to one of the problems (3.13), (3.14) and (3.16). Further, the same candidate PVs are considered for all CVs. Then, the number of candidate CVs (or candidate PV subsets) is $n_C = C^{n_y}_{n_{\text{cs}}}$. From this set, n_u CVs must be selected whereas order is irrelevant and multiple selections are allowed for covering also column structures. Thus, the number of elements in the common/individually-sized superstructure is $C^{n_C+n_u-1}_{n_u}$. For the dimensions $n_u = 3$, $n_y = 20$ and $n_{\text{cs}} = 4$, the number of elements in the superstructure yields $1.9 \cdot 10^{10}$. For each element, one of (3.13), (3.14) or (3.16) subject to its structure must be solved. Assuming exhaustive enumeration and 0.1 s expense per element, this means a total computation time of approximately 60 years.

It can be concluded that problems (3.13), (3.14) and (3.16) subject to common/individually-sized superstructures state mixed-integer problems in which global optimality per element is costly and the number of integer variants is potentially huge. Both aspects lead to large computational load for practical relevant problems. This can be tackled by reducing the number of evaluations via the use of BAB algorithms and by increasing the efficiency of iterative solution techniques. A further strategy is to drop the requirement of global optimality as a compromise for achieving better computational efficiency. In the next section, problems (3.13) and (3.16) subject to common/individually-sized superstructures are transformed into pure integer problems permitting suboptimality. These integer problems can then be solved very efficiently using BAB methods.

3.4.4. Space transformation

In order to account for the possibility of individual PV subsets $\mathcal{Y}_i$, (3.11) is restated as

$$\boldsymbol{z}^T \sum_{i=1}^{n_u} (\boldsymbol{G}_{\boldsymbol{z}}^{\boldsymbol{y}})^T \, \boldsymbol{h}_i \, \boldsymbol{e}_i^T = \boldsymbol{f}^T \sum_{i=1}^{n_u} \tilde{\boldsymbol{F}}^T \, \boldsymbol{h}_i \, \boldsymbol{e}_i^T, \tag{3.23}$$

where $\boldsymbol{h}_i^T \in \mathbb{R}^{n_y}$ is the ith row vector of $\boldsymbol{H}$, *i.e.*, $\boldsymbol{H}^T = \begin{bmatrix} \boldsymbol{h}_1 & \dots & \boldsymbol{h}_{n_u} \end{bmatrix}$, and $\boldsymbol{e}_i \in \mathbb{R}^{n_u}$ is the ith standard basis vector. The vector $\boldsymbol{h}_i = \begin{bmatrix} h_{i1} & \dots & h_{in_y} \end{bmatrix}^T$ represents the map from the complete PV set towards the ith CV. Considering a certain element of a common/individually-sized superstructure, a linear combination of the PV subset $\mathcal{Y}_i$ is mapped towards the ith CV. As $h_{ij} = 0 \, \forall j \notin \mathcal{Y}_i$, it is convenient to write

$$\boldsymbol{z}^T \sum_{i=1}^{n_u} (\boldsymbol{G}_{\boldsymbol{z}}^{\mathcal{Y}_i})^T \, \boldsymbol{h}_{i\mathcal{Y}_i} \, \boldsymbol{e}_i^T = \boldsymbol{f}^T \sum_{i=1}^{n_u} \left(\tilde{\boldsymbol{F}}^{\mathcal{Y}_i}\right)^T \, \boldsymbol{h}_{i\mathcal{Y}_i} \, \boldsymbol{e}_i^T, \tag{3.24}$$

where the sub- and superscript $\mathcal{Y}_i$ denotes that those rows of $\boldsymbol{G}_{\boldsymbol{z}}^{\boldsymbol{y}}$ and $\tilde{\boldsymbol{F}}$ and elements of $\boldsymbol{h}_i$ are extracted whose index is in $\mathcal{Y}_i$. *I.e.*, $\boldsymbol{h}_{i\mathcal{Y}_i}$ represents the map from the PVs of the subset $\mathcal{Y}_i$ towards the ith CV.

By performing the GSVD of the matrix pair $\left\{ \left(\tilde{\boldsymbol{F}}^{\mathcal{Y}_i}\right)^T, (\boldsymbol{G}_{\boldsymbol{z}}^{\mathcal{Y}_i})^T \right\}$ with

$$q_i = \operatorname{rank}\left(\begin{bmatrix} \tilde{\boldsymbol{F}}^{\mathcal{Y}_i} & \boldsymbol{G}_{\boldsymbol{z}}^{\mathcal{Y}_i} \end{bmatrix} \right),$$

(3.24) can be written as

$$\boldsymbol{z}^T \sum_{i=1}^{n_u} \boldsymbol{U}_i \boldsymbol{\Sigma}_i \tilde{\boldsymbol{h}}_i \boldsymbol{e}_i^T = \boldsymbol{f}^T \sum_{i=1}^{n_u} \tilde{\boldsymbol{U}}_i \tilde{\boldsymbol{\Sigma}}_i \tilde{\boldsymbol{h}}_i \boldsymbol{e}_i^T, \tag{3.25}$$

where

$$\boldsymbol{\Sigma}_i = \begin{bmatrix} \boldsymbol{0}_{t_i \times r_i} & \\ & \operatorname{diag}\left(\beta_{i(r_i+1)}, ..., \beta_{iq_i}\right) \end{bmatrix}, \tag{3.26}$$

$r_i = q_i - \operatorname{rank}\left(\boldsymbol{G}_{\boldsymbol{z}}^{\mathcal{Y}_i}\right)$, $t_i = n_{\boldsymbol{u}} - \operatorname{rank}\left(\boldsymbol{G}_{\boldsymbol{z}}^{\mathcal{Y}_i}\right)$,

$$\tilde{\boldsymbol{\Sigma}}_i = \begin{bmatrix} \operatorname{diag}\left(\alpha_{i1}, ..., \alpha_{i(q_i-s_i)}\right) & \\ & \boldsymbol{0}_{u_i \times s_i} \end{bmatrix},$$

$s_i = q_i - \operatorname{rank}\left(\tilde{\boldsymbol{F}}^{\mathcal{Y}_i}\right)$, $u_i = n_{\boldsymbol{f}} - \operatorname{rank}\left(\tilde{\boldsymbol{F}}^{\mathcal{Y}_i}\right)$; $\boldsymbol{U}_i \in \mathbb{R}^{n_{\boldsymbol{u}} \times n_{\boldsymbol{u}}}$ and $\tilde{\boldsymbol{U}}_i \in \mathbb{R}^{n_{\boldsymbol{f}} \times n_{\boldsymbol{f}}}$ are unitary; and $\tilde{\boldsymbol{h}}_i = \boldsymbol{V}_i^T \boldsymbol{h}_{i\mathcal{Y}_i}$, with $\boldsymbol{V}_i \in \mathbb{R}^{n_{\mathcal{Y}_i} \times q_i}$ and $\operatorname{rank}(\boldsymbol{V}_i) = q_i$.

The solution space of a single predefined common/individually-sized structure is now transformed. The task is to find the best $\tilde{\boldsymbol{h}}_i$ in terms of least worst-case/average loss and applying the back transformation $\boldsymbol{h}_{i\mathcal{Y}_i} = \boldsymbol{V}_i^{-T} \tilde{\boldsymbol{h}}_i$ if $q_i = n_{\mathcal{Y}_i}$ and otherwise $\boldsymbol{h}_{i\mathcal{Y}_i} = \left(\boldsymbol{V}_i^T\right)^{\dagger} \tilde{\boldsymbol{h}}_i$ in order to conclude to $\boldsymbol{H}$. It is important to stress that due to implications of Remark 3.10, $\left\|\tilde{\boldsymbol{h}}_i\right\|_2$ and $\|\boldsymbol{h}_{i\mathcal{Y}_i}\|_2$ can be chosen to be an arbitrary positive value.

The first useful property of the decomposed formulation (3.25) is that it allows the reduction of the solution space for one element of the common/individually-sized superstructure provided that suboptimality of the solution is permitted. This is clarified next.

Remark 3.13. From (3.26) it follows that if $r_i > 0$ for any $i \in \{1, \dots, n_{\boldsymbol{u}}\}$, then $\beta_{ij} = 0$, $\alpha_{ij} = 1 \forall j \in \{1, \dots, r_i\}$. *I.e.*, selecting the first r_i elements in $\tilde{\boldsymbol{h}}_i$ unequal zero corresponds to the unfavorable selection of infinite generalized singular values. Note that if the first r_i elements are the only nonzero elements in $\tilde{\boldsymbol{h}}_i$, then $\boldsymbol{h}_i$ is in the nullspace of $\left(\boldsymbol{G}_{\boldsymbol{z}}^{\mathcal{Y}_i}\right)^T$ and $\sum_{i=1}^{n_u} \left(\boldsymbol{G}_{\boldsymbol{z}}^{\mathcal{Y}}\right)^T \boldsymbol{h}_i \boldsymbol{e}_i^T$ is singular which is strictly undesirable. Thus, it is generally advisable to select only the last $q_i - r_i$ elements in $\tilde{\boldsymbol{h}}_i$ unequal zero. Accordingly, the solution space can be reduced from $\sum_{i=1}^{n_u} n_{\mathcal{Y}_i}$ to $\sum_{i=1}^{n_u} q_i - r_i = \sum_{i=1}^{n_u} \operatorname{rank}\left(\boldsymbol{G}_{\boldsymbol{z}}^{\mathcal{Y}_i}\right) \leq n_{\boldsymbol{u}}^2$.

Based on the transformed solution space for one element, a novel method for solving the worst-case/average loss problem (3.13)/(3.16) subject to common/individually-sized superstructures is presented in the subsequent section.

3.4.5. AM problem solution

In the case of a large number of elements in a common/individually-sized superstructure, it is not affordable to perform costly iterative solution approaches for each element. A good compromise between optimality and efficiency is obtained by reducing the solution space to discrete points. Considering (3.25), *i.e.*, the map from $\boldsymbol{f}$ to $\boldsymbol{z}$ for a certain element of the superstructure in terms of the transformed space, the limitation $\tilde{\boldsymbol{h}}_i \in \{\tilde{\boldsymbol{e}}_{r_i+1}, \ldots, \tilde{\boldsymbol{e}}_{q_i}\}$ is imposed for every CV, where $\tilde{\boldsymbol{e}}_j \in \mathbb{R}^{q_i}$ refers to the jth basis vector. Take into account that $\tilde{\boldsymbol{h}}_i \notin \{\tilde{\boldsymbol{e}}_1, \ldots, \tilde{\boldsymbol{e}}_{r_i}\}$ is due to Remark 3.13. Now, finding the solution of the worst-case/average loss problem (3.13)/(3.16) subject to a predefined common/individually--sized control structure only requires the screening of $\prod_{i=1}^{n_u} q_i - r_i$ alternatives instead of solving a nonconvex problem with $\sum_{i=1}^{n_u} q_i - r_i$ variables. The more striking advantage of this approach is that the problem of finding the best element in a common/individually-sized superstructure is transformed from a mixed-integer into a pure integer problem and the integer problem per element can be combined with the integer problem of finding the element in the superstructure.

For the jth candidate PV subset of the ith CV a GSVD must be performed yielding the matrices $\boldsymbol{U}_i^{(j)}$, $\boldsymbol{\Sigma}_i^{(j)}$, $\tilde{\boldsymbol{U}}_i^{(j)}$, $\tilde{\boldsymbol{\Sigma}}_i^{(j)}$. The problem of finding the best linear combination of the best candidate PV subset of the ith CV now corresponds to finding the best entry among all entries of $\tilde{\boldsymbol{h}}_i^{(j)} \forall j \in \{1, \ldots, n_C\}$. This in turn corresponds to finding the best relating columns of the augmented matrices $\boldsymbol{\mathcal{G}}_i$ and $\boldsymbol{\mathcal{F}}_i$ given by

$$\boldsymbol{\mathcal{G}}_i^T = \begin{bmatrix} \boldsymbol{U}_i^{(1)} \boldsymbol{\Sigma}_i^{(1)\diamond} & \ldots & \boldsymbol{U}_i^{(nc)} \boldsymbol{\Sigma}_i^{(nc)\diamond} \end{bmatrix}$$
$$\boldsymbol{\mathcal{F}}_i^T = \begin{bmatrix} \tilde{\boldsymbol{U}}_i^{(1)} \tilde{\boldsymbol{\Sigma}}_i^{(1)\diamond} & \ldots & \tilde{\boldsymbol{U}}_i^{(nc)} \tilde{\boldsymbol{\Sigma}}_i^{(nc)\diamond} \end{bmatrix}.$$

The superscript $\diamond$ denotes that the first r_j columns in $\boldsymbol{\Sigma}_i^{(j)}$ and $\tilde{\boldsymbol{\Sigma}}_i^{(j)}$ are omitted which is due to Remark 3.13. Accordingly, $\boldsymbol{\mathcal{G}}_i \in \mathbb{R}^{v \times n_u}$ and $\boldsymbol{\mathcal{F}}_i \in \mathbb{R}^{v \times n_f}$ where $v = \sum_{j=1}^{nc} q_j - r_j$.

It was already pointed out in Remark 3.11 that in this thesis for every CV the same set of candidates PV subsets are taken into account. Thus for every CV the augmented matrices are identical and index i can be dropped. Now, the map from $\boldsymbol{f}$ to $\boldsymbol{z}$ can be written in terms of all elements of the superstructure as

$$\boldsymbol{z}^T \boldsymbol{\mathcal{G}}^T \boldsymbol{\mathcal{H}}^T = \boldsymbol{f}^T \boldsymbol{\mathcal{F}}^T \boldsymbol{\mathcal{H}}^T. \tag{3.27}$$

and the problem of finding common-/individually-sized solutions is reduced to solving the problem

$$\boldsymbol{\mathcal{H}} = \arg\min_{\boldsymbol{\mathcal{H}}} \left\| (\boldsymbol{\mathcal{H}} \boldsymbol{\mathcal{G}})^{-1} \boldsymbol{\mathcal{H}} \boldsymbol{\mathcal{F}} \right\|_\star \text{ s.t. } \boldsymbol{\mathcal{H}}_{ij} \in \{0,1\} \text{ and } \boldsymbol{\mathcal{H}} \boldsymbol{\mathcal{H}}^T = \boldsymbol{I}_{n_u}. \tag{3.28}$$

Here, $\boldsymbol{\mathcal{H}} \in \mathbb{R}^{n_u \times v}$ is a binary matrix with one nonzero entry per column and $\star$ refers to either induced two-norm or Frobenius norm depending on whether worst-case or average loss is considered. Note that problem (3.28) is almost identical to the original problems

(3.13) and (3.16) subject to PV selection. Analogously to (3.14), an approximate solution can be obtained by solving an MSV problem

$$\mathcal{H} = \arg\max_{\mathcal{H}} \underline{\sigma}\left(\mathcal{H}\,\mathcal{G}\right) \text{ s.t. } \mathcal{H}_{ij} \in \{0,1\} \text{ and } \mathcal{H}\,\mathcal{H}^T = \boldsymbol{I}_{n_u}. \tag{3.29}$$

Problems (3.28) and (3.29) are referred to as augmented matrix (AM) problems. They are pure integer problems with a combinatorial complexity of $C_{n_u}^{v}$ integer solutions. As this may be a huge number also for rather small I/O dimensions, very efficient solutions methods for problems (3.28) and (3.29) are needed. Due to their similarity to the PV selection problems, very efficient methods are already available in form of the bidirectional BAB (B3) algorithms proposed by Cao and Kariwala (2008) and Kariwala and Cao (2009, 2010a). They are outlined in the sequel.

B3MSV This method is based on the work of Cao and Kariwala (2008). A MATLAB® code was made publicly available on the Internet by Cao (2007/11/11). The method solves the MSV problem (3.14) subject to PV selection, *i.e.*, it selects the rows of a tall matrix such that the resulting square matrix has maximum MSV.

B3WC, PB3WC These methods were developed by Kariwala and Cao (2009) and made publicly available as MATLAB® codes by Cao (2009/01/09). The B3WC method solves the worst-case loss problem (3.13) subject to PV selection. The PB3WC solves the worst-case loss problem (3.13) subject to a column superstructure. As it is not fully bidirectional, it is called partial B3 algorithm.

B3AV, PB3AV These methods were developed by Kariwala and Cao (2010a) and made publicly available as MATLAB® codes by Cao (2009/11/17). The B3AV method solves the average loss problem (3.16) subject to PV selection. The PB3AV solves the average loss problem (3.16) subject to a column superstructure. As the PB3WC, the PB3AV method is based on a partial B3 algorithm.

The application of the B3MSV algorithm to (3.29) is referred to as the B3MSV-AM method and can take place without modifications. The B3WC and B3AV algorithm need to be slightly modified to be applicable to problem (3.28). The necessary changes in the algorithm are presented in Appendix 3.C. The use of either of the modified algorithms for solving (3.28) is referred to as B3WC-AM and B3AV-AM method.

A preprocessing step of all AM methods is the calculation of the augmented matrices. If common-sized control structures with set size n_{cs} are desired, then the preprocessing step involves $n_{\mathcal{C}} = C_{n_{\text{cs}}}^{n_y}$ GSVD calls. On the one hand this may already be a computational expensive task for large I/O dimensions. On the other hand the B3 algorithms rely on calculating $v \times v$ matrices which limits $n_{\mathcal{C}}$ due to memory reasons. Additional constraints on the candidate PV subsets such as disregarding mixed-unit combinations may reduce the size of n_C and v. An additional possibility for keeping the computational load small is to predefine candidate PV subsets from process insight.

3.5. Numerical study

Five CSD methods were tested against each other using MATLAB®, the PB3WC, PB3WC-LT, B3MSV-AM, the B3WC-AM and the MIWC method. The PB3WC-LT and the MIWC method have not been introduced yet.

PB3WC-LT This method refers to finding a common-sized control structure of PV set size n_{cs} by applying the PB3WC method to find the best column structure of PV set size $n_{\text{cs}} + n_{\boldsymbol{u}} + 1$ and then linearly transforming the result into a n_{cs} sized common-sized structure.

MIWC/MIAV This method refers to the approximate solution of the worst-case/average loss problem (3.13)/(3.16) subject to the common/individually-sized superstructure, which is of mixed-integer type. For each element of the superstructure the decomposition (3.25) is taken into account for the sake of searching in the reduced space and having good starting values, *i.e.*, $\tilde{\boldsymbol{h}}_i = \tilde{\boldsymbol{e}}_{q_i} \, \forall i \in \{1, \ldots, n_{\boldsymbol{u}}\}$. The problem per element is locally solved by unconstrained minimization (MATLAB® function fminunc()) based on finite-difference approximation of derivatives. The computational efficiency of the integer problem is enhanced by BAB algorithms as clarified in Appendix 3.D.

All methods were applied to several cases whereas each case was 1000 times repeatedly calculated with randomly generated linear state-space models (3.2), scaling matrices $\boldsymbol{W_d}$ and $\boldsymbol{W_{n^y}}$ as well as objectives of the form (3.3). The entries of the coefficient matrices followed a normal distribution with $\mathcal{N}(0,1)$ except for the uniformly distributed $\boldsymbol{W_d} \sim \mathcal{U}(0,1)$ and $\boldsymbol{W_{n^y}} \sim \mathcal{U}(0, 1.0E-4)$ as well as $J_{\boldsymbol{uu}} = \boldsymbol{R}\boldsymbol{R}^T$ where the coefficients of $\boldsymbol{R} \in \mathbb{R}^{n_u \times n_u}$ were normally distributed with $\mathcal{N}(0,1)$ and the bound $\underline{\lambda}(J_{\boldsymbol{uu}}) > 1.0 \cdot 10^{-4}$ was imposed. An alternative measure of optimality, the relative loss $\Lambda_\star$, was used. $\Lambda_\star$ specifies the ratio between the loss of the result of a particular method and the least loss of all PV selection structures. The subscript $\star$ relates to either 2 or F depending on whether worst-case or average loss is considered. The $\Lambda_\star$ measure is convenient as the loss of PV selection represents the upper bound for the loss of PV combination, *i.e.*, $\Lambda_2 \leq 1$ if worst-case loss is considered.

Two test runs were performed. In the first run (A), eight different cases were simulated. From case to case, the number of combined PVs $n_{\mathcal{Y}}$ was decreased from eight to one,

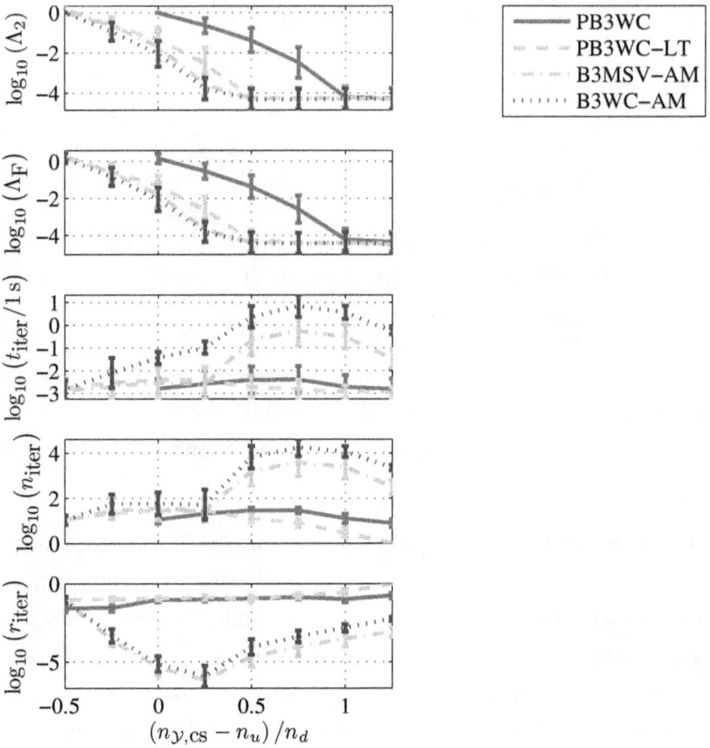

Figure 3.3.: Numerical results of run A

whereas the other dimensions, $n_u = 3$, $n_d = 4$ and $n_y = 10$, remained unaltered. In the second run (B), $n_{\mathcal{Y}}$ was reduced from six to one in six steps while keeping $n_u = 2$, $n_d = 3$ and $n_y = 7$ fixed. The calculations were conducted in MATLAB® R2008b using a Windows XP® SP2 desktop with Intel® Core™ Duo CPU E8400 (3.0 GHz, 3.5 GB RAM).

Figure 3.3 shows the results of run A. The following measures are presented in the charts from top to bottom: relative worst-case and average loss, Λ_2 and Λ_F, computation time t_{iter}, evaluation number n_{iter} and relative evaluation number r_{iter} which represents the evaluation number related to the number of elements in the superstructure. The ordinate values are all indicated in terms of mean values with standard deviation bar. For the abscissa, the relative measure for the PV subset size $\nu = (n_{\mathcal{Y},cs} - n_u)/n_d$ is used, which is convenient as the characteristic points $\nu = 0$ and $\nu = 1$ relate to $n_{\mathcal{Y},cs} = n_u$ and $n_{\mathcal{Y},cs} = n_u + n_d$, respectively.

The following observations can be made. In agreement with Lemma 3.7, the loss curves of the PB3WC-LT are identical with those of the PB3WC but shifted by $\Delta n_{\mathcal{Y},cs} = n_u - 1$,

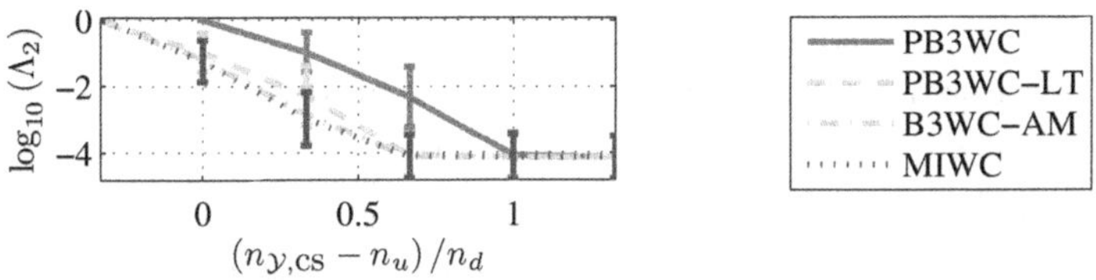

Figure 3.4.: Numerical results of run B

i.e., $\Delta\nu = (n_{\boldsymbol{u}} - 1)/n_{\boldsymbol{d}}$, to the left. The PB3WC fails for $\nu < 0$ and achieves equal worst-case loss as PV selection for $\nu = 0$, which is clear in the context of Corollary 3.8. The mean relative worst-case/average loss of each method monotonically declines with increasing ν. However, at some points, a lower bound for the worst-case/average loss is reached where a significant reduction cannot be achieved anymore. This lower bound of the loss is achieved by the optimal solution to the worst-case loss problem (3.13) subject to the combination of the complete PV subset and is called best disturbance rejection loss (BDRL). From theory it is evident that the BDRL is zero if the implementation error is disregarded and $n_{\boldsymbol{y}} \geq n_{\boldsymbol{u}} + n_{\boldsymbol{d}}$. The BDRL is almost achieved by the PB3WC method at $\nu = 1$ and by the PB3WC-LT and the other methods at $\nu = 1 - \Delta\nu$. It means that the selection $\nu > 1 - \Delta\nu$, *i.e.*, $n_{\mathcal{Y},\mathrm{cs}} > n_{\boldsymbol{d}} + 1$, is unreasonable for common/individually-sized control structures as the achievable worst-case/average loss reduction to the case $\nu = 1 - \Delta\nu$ is insignificant. For $\nu \geq 1 - \Delta\nu$, the computational load of both the B3WC-AM and the B3MSV-AM method is considerably larger than that of the other methods making them unreasonable in terms of achievable loss reduction. The excellence of the B3 methods is indicated in the lowermost chart. Related to the combinatorial complexity the B3 methods are able to reduce the number of evaluations up to 6 orders of magnitude. As expected, the B3MSV-AM method shows better computational efficiency than the B3WC-AM method as the latter is not fully bidirectional.

The main result of run B is presented in Figure 3.4. It illustrates that the relative worst-case loss of the MIWC method is smaller than but almost indistinguishable close to that of the B3WC-AM. This justifies that the simplifications made for the AM methods do not severely impair the achievable accuracy of the solution.

3.6. Conclusions

This chapter gives new insights into the problem of finding self-optimizing control structures. Several new methods have been proposed. The GSVD method is dedicated to finding proper CVs, altogether linear combinations of a common PV subset. As the methods by Kariwala *et al.* (2008) and Alstad *et al.* (2009), it minimizes the average loss superoptimal to the worst-case loss by taking expected disturbances and measurement errors into account. It is also the basis for the newly developed AM methods which are able to find suboptimal control structures with the structural constraint that for each CV individual PV subsets can be taken into account. As no restriction on the size of the PV subset was made, the AM methods close the link between PV selection and combination. The AM methods proved their usefulness by the application to randomly generated systems and practical examples.

A major assumption for the derivation of the AM methods is that for every CV the same candidate PV subset needs to be taken into account. As it is desirable to relax this constraint, future work may aspire the further development of the AM methods. Alternatively, the B3 algorithms (Cao and Kariwala, 2008; Kariwala and Cao, 2009, 2010a) may be enhanced such that every CV can be related to a candidate PV subset and structural solutions which violate this relation are pruned.

3.A. PV combination methods

The list below presents solution methods of problem (3.13) and (3.16) subject to PV combination of a common PV subset per CV. All methods are written in terms of the complete PV set $\mathcal{Y}_{\boldsymbol{y}}$. Note that they work as well for a selected PV subset $\mathcal{Y} \subseteq \mathcal{Y}_{\boldsymbol{y}}$. Then, in the formulas stated below the respective rows in $\boldsymbol{G^y}$ and $\boldsymbol{G_d^y}$ must be extracted and $n_{\boldsymbol{y}}$ must be substituted by $n_{\mathcal{Y}} = |\mathcal{Y}|$.

Null space method Alstad and Skogestad (2007) proposed a simple method for the identification of PV combinations. They required that $\boldsymbol{H}$ is the left null space (kernel) of

$$\boldsymbol{F} = \boldsymbol{G_d^y} - \boldsymbol{G^y}\, J_{\boldsymbol{uu}}^{-1}\, J_{\boldsymbol{ud}},$$

represented by the notation $\boldsymbol{H}^T = \ker\left(\boldsymbol{F}^T\right)$. This selection provides

$$\boldsymbol{H}\,\boldsymbol{F} = \boldsymbol{0}, \tag{3.30}$$

which in turn makes the loss independent of the disturbances as (3.10c) and (3.11) reveals. The necessary condition for the application for this method is $n_{\boldsymbol{y}} \geq n_{\boldsymbol{u}} + n_{\boldsymbol{d}}$. If the inequality holds, a preselection of a subset of PVs $\mathcal{Y}$ has to be performed such that the subset size equals $n_{\boldsymbol{u}} + n_{\boldsymbol{d}}$. One drawback of the method is that this may be an unfavorable large number. Another drawback is due to the fact that the implementation errors $\boldsymbol{n^y}$ are neglected.

Extended null space method The null space method has been extended by Alstad *et al.* (2009) to the general case with extra measurements, *i.e.*, $n_{\boldsymbol{y}} > n_{\boldsymbol{u}} + n_{\boldsymbol{d}}$, and to few measurements, *i.e.*, $n_y < n_u + n_d$. The idea of the extended nullspace method is to first focus on minimizing the steady-state loss caused by disturbances, and then, if there are remaining degrees of freedom, minimize the effect of measurement errors. The method is based on the fact that $\boldsymbol{M}(\boldsymbol{H}) = \boldsymbol{M}(\boldsymbol{M}_n \boldsymbol{H})$ provided $\boldsymbol{M}_n \in \mathbb{R}^{n_u \times n_u}$ is regular. By setting $\boldsymbol{M}_n = (\boldsymbol{H}\,\boldsymbol{G}^y_{\boldsymbol{z}})^{-1}$ and $\boldsymbol{W}_{n^y} = \boldsymbol{0}$, (3.11) can be restated as

$$\boldsymbol{z} = \boldsymbol{M}_n \boldsymbol{H} \underbrace{\begin{bmatrix} \boldsymbol{G}^y & \boldsymbol{G}^y_d \end{bmatrix}}_{=\tilde{\boldsymbol{G}}} \begin{bmatrix} \boldsymbol{u}_{\mathrm{R/O}} \\ \boldsymbol{d} \end{bmatrix}. \tag{3.31}$$

For feed-forward control, the loss variables $\boldsymbol{z}$ are given by

$$\boldsymbol{z} = \tilde{\boldsymbol{J}} \begin{bmatrix} \boldsymbol{u} \\ \boldsymbol{d} \end{bmatrix}, \tag{3.32}$$

where

$$\tilde{\boldsymbol{J}} = \begin{bmatrix} \boldsymbol{J}_{\boldsymbol{uu}}^{1/2} & \boldsymbol{J}_{\boldsymbol{uu}}^{1/2} \boldsymbol{J}_{\boldsymbol{uu}}^{-1} \boldsymbol{J}_{\boldsymbol{ud}} \end{bmatrix}.$$

By comparison of (3.31) and (3.32), it follows that the solution of

$$\boldsymbol{M}_n \boldsymbol{H} \tilde{\boldsymbol{G}} = \tilde{\boldsymbol{J}} \tag{3.33}$$

gives the desired $\boldsymbol{H}$. For the cases $n_y \geq n_u + n_d$ the solution of (3.33) gives zero disturbance loss, *i.e.*, $\|\boldsymbol{z}\|_2 = 0$. If $n_y > n_u + n_d$, the left pseudo-inverse of $\tilde{\boldsymbol{G}}$ can be used to determine $\boldsymbol{H}$. If $n_y < n_u + n_d$, the right pseudo-inverse of $\tilde{\boldsymbol{G}}$ can be used to determine $\boldsymbol{H}$, minimizing $\|\boldsymbol{E}\|_F$ in which $\boldsymbol{E} = \boldsymbol{M}_n \boldsymbol{H} \tilde{\boldsymbol{G}} - \tilde{\boldsymbol{J}}$. Alstad *et al.* (2009) prove that this also minimizes $\|\boldsymbol{M}\|_{\mathrm{F}}$. To account for the implementation errors the authors proposed to select

$$\boldsymbol{H} = \boldsymbol{M}_n^{-1} \tilde{\boldsymbol{J}} \left(\boldsymbol{W}_{n^y}^{-1} \tilde{\boldsymbol{G}}\right)^{\dagger} \boldsymbol{W}_{n^y}^{-1},$$

where $\dagger$ indicates the pseudo-inverse of a matrix. They proved that this selection minimizes both $\|\boldsymbol{E}\|_F$ and $\|\boldsymbol{M}_n \boldsymbol{H} \boldsymbol{W}_{n^y}\|_{\mathrm{F}}$ and pointed out that the latter norm may be interpreted as the effect of the implementation error on the loss variables $\boldsymbol{z}$.

Constrained average loss minimization Alstad *et al.* (2009) published another method, also based on the principle that the problems (3.13)/ (3.16) have no unique solution. From this fact the authors conclude that it is useful to select $\boldsymbol{M}_n = (\boldsymbol{H}\,\boldsymbol{G}^y_{\boldsymbol{z}})^{-1} = \boldsymbol{I}_{n_u}$ and minimize (3.16) subject to this constraint. The optimization problem can then be restated as

$$\boldsymbol{H} = \arg\min_{\boldsymbol{H}^T} \left\| \tilde{\boldsymbol{F}}^T \boldsymbol{H}^T \right\|_F \quad \text{s.t.} \quad (\boldsymbol{G}^y)^T \boldsymbol{H}^T = \left(J_{\boldsymbol{uu}}^{1/2}\right)^T.$$

This linearly constrained convex problem has the explicit solution

$$\boldsymbol{H}^T = \left(\tilde{\boldsymbol{F}}\tilde{\boldsymbol{F}}^T\right)^{-1} \boldsymbol{G}^y \left((\boldsymbol{G}^y)^T \left(\tilde{\boldsymbol{F}}\tilde{\boldsymbol{F}}^T\right)^{-1} \boldsymbol{G}^y\right)^{-1} \left(J_{\boldsymbol{uu}}^{1/2}\right)^T. \tag{3.34}$$

Note that no minimization takes place for the cases in which $n_{\boldsymbol{y}} = n_{\boldsymbol{u}}$. This is due to the fact that the solution of the constraint is unique, leaving no degree of freedom for the minimization. The optimality is then exclusively a matter of selecting an appropriate subset $\mathcal{Y}$. A generalization thereof is pointed out in Corollary 3.8. In the following, it is shown that the solution (3.34) provides the same worst-case/average loss as the method by Kariwala *et al.* (2008). Inserting (3.34) into (3.10c) yields

$$\boldsymbol{M} = J_{\boldsymbol{uu}}^{1/2} \left((\boldsymbol{G}^y)^T \left(\tilde{\boldsymbol{F}}\tilde{\boldsymbol{F}}^T\right)^{-1} \boldsymbol{G}^y\right)^{-1} (\boldsymbol{G}^y)^T \left(\tilde{\boldsymbol{F}}\tilde{\boldsymbol{F}}^T\right)^{-1} \tilde{\boldsymbol{F}}.$$

The average loss (3.15) is then given by

$$\begin{aligned} 6\,(n_{\boldsymbol{y}} + n_{\boldsymbol{d}})\, L_{\text{av}} = \|\boldsymbol{M}\|_{\text{F}}^2 = \operatorname{tr}\left(\boldsymbol{M}\,\boldsymbol{M}^T\right) &= \sum_{i=1}^{n_u} \lambda_i \left(\boldsymbol{M}\,\boldsymbol{M}^T\right) \\ &= \sum_{i=1}^{n_u} \lambda_i \left(J_{\boldsymbol{uu}}^{1/2} \left((\boldsymbol{G}^y)^T \left(\tilde{\boldsymbol{F}}\tilde{\boldsymbol{F}}^T\right)^{-1} \boldsymbol{G}^y\right)^{-1} \left(J_{\boldsymbol{uu}}^{1/2}\right)^T\right), \\ &= \sum_{i=1}^{n_u} \lambda_i^{-1} \left(J_{\boldsymbol{uu}}^{-1/2} (\boldsymbol{G}^y)^T \left(\tilde{\boldsymbol{F}}\tilde{\boldsymbol{F}}^T\right)^{-1} \boldsymbol{G}^y \left(J_{\boldsymbol{uu}}^{-1/2}\right)^T\right), \end{aligned}$$

which agrees with (3.36), the average loss by Kariwala *et al.* (2008). Note that λ_i indicates the ith largest eigenvalue.

Minimum worst-case loss by eigenstructure analysis Kariwala (2007) proved that the following algorithm leads to a global solution of (3.13) subject to a common PV subset per CV.

1. Perform the singular value decomposition

$$\begin{bmatrix} \boldsymbol{G}^y J_{\boldsymbol{uu}}^{-1/2} & \tilde{\boldsymbol{F}} \end{bmatrix} = \boldsymbol{U} \begin{bmatrix} \boldsymbol{\Sigma} & \boldsymbol{0} \end{bmatrix} \begin{bmatrix} \boldsymbol{V}_{11} & \boldsymbol{V}_{12} \\ \boldsymbol{V}_{21} & \boldsymbol{V}_{22} \end{bmatrix}^T$$

   such that $\boldsymbol{G}^y J_{\boldsymbol{uu}}^{-1/2} = \boldsymbol{U}\,\boldsymbol{\Sigma}\,\boldsymbol{V}_{11}^T$.

2. Calculate $\gamma = \sqrt{\frac{1}{\sigma_{n_u}^2(\boldsymbol{V}_{11})} - 1}$, where σ_i denotes the ith largest singular value. Then $\lambda_{n_u}\left(\gamma^2\, \boldsymbol{G}^y\, J_{\boldsymbol{uu}}^{-1}\, (\boldsymbol{G}^y)^T - \tilde{\boldsymbol{F}}\,\tilde{\boldsymbol{F}}^T\right) = 0$, where λ_i indicates the ith largest eigenvalue.

3. Perform an eigenvalue decomposition of $\gamma^2\, \boldsymbol{G}^y\, J_{\boldsymbol{uu}}^{-1}\, (\boldsymbol{G}^y)^T - \tilde{\boldsymbol{F}}\,\tilde{\boldsymbol{F}}^T$ and find the eigenvectors $\boldsymbol{\nu}_1, \dots, \boldsymbol{\nu}_{n_u}$ corresponding to the largest n_u eigenvalues.

4. Select the PV combinations as $\boldsymbol{H} = \boldsymbol{M}_n \begin{bmatrix} \boldsymbol{\nu}_1 & \dots & \boldsymbol{\nu}_{n_u} \end{bmatrix}^T$, where $\boldsymbol{M}_n \in \mathbb{R}^{n_u \times n_u}$ can be any regular matrix.

The minimum worst-case loss is then given by

$$L_{\text{wc}} = \frac{1}{2}\gamma^2. \tag{3.35}$$

Minimum average loss by eigenstructure analysis Kariwala *et al.* (2008) proved that the following algorithm leads to a global optimal solution to (3.16) subject to a common PV subset per CV.

1. Evaluate $\boldsymbol{X} = J_{\boldsymbol{uu}}^{-1/2}\, (\boldsymbol{G}^y)^T \left(\tilde{\boldsymbol{F}}\,\tilde{\boldsymbol{F}}^T\right)^{-1} \boldsymbol{G}^y\, J_{\boldsymbol{uu}}^{-1/2}$.

2. Perform an eigenvalue decomposition of $\boldsymbol{G}^y\, J_{\boldsymbol{uu}}^{-1/2}\, \boldsymbol{X}\, J_{\boldsymbol{uu}}^{-1/2}\, (\boldsymbol{G}^y)^T - \tilde{\boldsymbol{F}}\,\tilde{\boldsymbol{F}}^T$ and find the eigenvectors $\boldsymbol{\nu}_1, \dots, \boldsymbol{\nu}_{n_u}$ to the largest n_u eigenvalues.

3. Select the PV combinations as $\boldsymbol{H} = \boldsymbol{M}_n \begin{bmatrix} \boldsymbol{\nu}_1 & \dots & \boldsymbol{\nu}_{n_u} \end{bmatrix}^T$, where $\boldsymbol{M}_n \in \mathbb{R}^{n_u \times n_u}$ can be any regular matrix.

The minimum average loss is then given by

$$L_{\text{av}} = \frac{1}{6\,(n_d + n_y)} \sum_{i=1}^{n_u} \lambda_i^{-1}(\boldsymbol{X}). \tag{3.36}$$

As pointed out above, the same result can be achieved by the constrained average loss minimization by Alstad *et al.* (2009).

3.B. Equivalence of CSDs methods

Next, it is shown that the minimum average loss by Kariwala *et al.* (2008) and Alstad *et al.* (2009), *i.e.*, equation (3.36), can be converted into the average loss of the GSVD method (3.20).

$$
\begin{aligned}
&\frac{1}{6\left(n_d+n_y\right)} \sum_{i=1}^{n_u} \lambda_i^{-1}\left(\left(G_z^y\right)^T\left(\tilde{F}\tilde{F}^T\right)^{-1} G_z^y\right) \\
\stackrel{(1)}{=}&\frac{1}{6\left(n_d+n_y\right)} \sum_{i=1}^{n_u} \lambda_i^{-1}\left(\left(G_z^y\right)^T R^{-T} R^{-1} G_z^y\right) \\
\stackrel{(2)}{=}&\frac{1}{6\left(n_d+n_y\right)} \sum_{i=1}^{n_u} \lambda_i^{-1}\left(R^{-1} G_z^y \left(G_z^y\right)^T R^{-T}\right) \\
\stackrel{(3)}{=}&\frac{1}{6\left(n_d+n_y\right)} \sum_{i=1}^{n_u} \lambda_i^{-1}\left(R^{-1} G_z^y \left(G_z^y\right)^T R^{-T}, I_{n_u}\right) \\
\stackrel{(4)}{=}&\frac{1}{6\left(n_d+n_y\right)} \sum_{i=1}^{n_u} \lambda_i^{-1}\left(G_z^y \left(G_z^y\right)^T, R R^T\right) \\
\stackrel{(5)}{=}&\frac{1}{6\left(n_d+n_y\right)} \sum_{i=n_y-n_u+1}^{n_y} \lambda_i\left(\tilde{F}\tilde{F}^T, G_z^y \left(G_z^y\right)^T\right) \\
\stackrel{(6)}{=}&\frac{1}{6\left(n_d+n_y\right)} \sum_{i=n_y-n_u+1}^{n_y} \sigma_i^2\left(\tilde{F}^T, \left(G_z^y\right)^T\right)
\end{aligned}
$$

In the first step, a Cholesky factorization of the matrix $\tilde{F}\tilde{F}^T = R R^T$ is performed. For the second step it must be proven that $\lambda_i\left(A A^T\right) = \lambda_i\left(A^T A\right) \forall i \leq m$ for any matrix $A \in \mathbb{R}^{m\times n} \mid m \leq n$. The evidence is trivial when considering the singular value decomposition $A = U_A \Sigma_A V_A^T$ with $\Sigma_A \Sigma_A^T = \operatorname{diag}\left(s_1^2, \ldots, s_m^2\right)$ and $\Sigma_A^T \Sigma_A = \operatorname{diag}\left(s_1^2, \ldots, s_m^2, 0, \ldots\right)$. In the third step the fact is used that every eigenvalue problem $\lambda(B)$, *i.e.*, $B x = \lambda x$ with $B \in \mathbb{R}^{m\times m}$, may be written as a generalized eigenvalue problem $\lambda(B, I_m)$, *i.e.*, $B x = \lambda I_m x$. In the textbook of Golub and VanLoan (1996, p. 461) it is pointed out that $\lambda(B, C) = \lambda\left(X^T B X, X^T C X\right)$ holds for symmetric B, $C \in \mathbb{R}^{m\times m}$ where $C \succ 0$ and any regular $X \in \mathbb{R}^{m\times m}$. Thereby, step four is justified. Step five includes the re-composition of the Cholesky factors and the interchange of the matrices in the arguments of the generalized eigenvalues. Considering the equation $B x = \lambda C x$ and $\lambda B x = C x$ for B, $C \in \mathbb{R}^{m\times m}$, it is clear that $\lambda_i(B, C) = \lambda_{m+1-i}^{-1}(C, B)$. Step six is clarified by the statement of Golub and VanLoan (1996, p. 466) that $\sigma^2(D, E) \in \lambda\left(D^T D, E^T E\right)$ for $D \in \mathbb{R}^{k\times n}$, $n \leq k$, and $E \in \mathbb{R}^{m\times n}$, $m \leq k$. For comparison with (3.20), it must be taken into account that $q = \operatorname{rank}\left(\tilde{F}\right) = n_y$ and $r = q - \operatorname{rank}\left(G_z^y\right) = n_y - n_u$.

The GSVD method (and others, Kariwala *et al.*, 2008; Alstad *et al.*, 2009) minimizes the average loss superoptimal to the worst-case loss. Therefore, its result for the worst-case loss must equal that by Kariwala (2007) which derived a method for exclusive worst-case loss minimization. The conversion between the minimum worst-case loss by Kariwala

(2007) and (3.19) is not shown here for the sake of brevity. The conversion steps are similar to the ones stated above.

3.C. Modification of B3WC/B3AV method

The B3WC/B3AV algorithm is specially designed for solving (3.13)/(3.16) subject to PV selection. *I.e.*, the integer problem of selecting $n_{\boldsymbol{u}}$ corresponding rows in $\boldsymbol{G}_{\boldsymbol{z}}^{\boldsymbol{y}}$ and $\tilde{\boldsymbol{F}}$ which minimize the worst-case/average loss must be solved. Instead of screening all possible solutions by exhaustive enumeration, the methods search in so called solution trees where only promising solutions and sub-solutions are evaluated. This is done by entering always the most promising path (branching) and exclude branches where the solution is know to be absent (pruning). One tree is in upward direction where the root node relates to an empty set of selected rows and where $n_{\boldsymbol{u}}$ node layers lie between root and branch ends. The other tree is in downward direction where the root node relates to the full set of rows selected and $n_{\boldsymbol{y}} - n_{\boldsymbol{u}}$ node layers are located in between root and branch ends. Both trees have $C_{n_{\boldsymbol{u}}}^{n_{\boldsymbol{y}}}$ branch ends which relate to the possible solutions. Searching in such trees makes it necessary to evaluate and judge intermediate nodes in which the number of selected rows is less or greater than $n_{\boldsymbol{u}}$. Further, from the evaluation of an intermediate node it must be possible to conclude whether further evaluations towards the end nodes are reasonable or not. For this purpose, Kariwala and Cao (2009, 2010a) introduced the functions

$$\mathcal{L}_{\text{wc}}(\mathcal{Y}) = \frac{1}{2}\lambda_w^{-1}(\boldsymbol{N}(\mathcal{Y})) \tag{3.37}$$

and

$$\mathcal{L}_{\text{av}}(\mathcal{Y}) = \sum_{i=1}^{w}\lambda_i^{-1}(\boldsymbol{N}(\mathcal{Y})), \tag{3.38}$$

respectively. Here, λ_i indicates the ith largest eigenvalue, $w = \text{rank}\left(\boldsymbol{G}_{\boldsymbol{z}}^{\mathcal{Y}}\right)$,

$$\boldsymbol{N}(\mathcal{Y}) = \left(\boldsymbol{G}_{\boldsymbol{z}}^{\mathcal{Y}}\right)^T\left(\tilde{\boldsymbol{F}}^{\mathcal{Y}}\left(\tilde{\boldsymbol{F}}^{\mathcal{Y}}\right)^T\right)^{-1}\boldsymbol{G}_{\boldsymbol{z}}^{\mathcal{Y}}$$

and $\mathcal{Y}$ indicates the selected PV subset of size $n_{\mathcal{Y}}$. Both functions, (3.37) and (3.38), can be evaluated for a $\mathcal{Y}$ with arbitrary set size, *i.e.*, $1 \leq n_{\mathcal{Y}} \leq n_{\boldsymbol{y}}$. They have the property that they are monotonically increasing from both empty and full subset towards a subset with size $n_{\boldsymbol{u}}$ along a certain branch. At set size $n_{\boldsymbol{u}}$, (3.37) and (3.38) equal the worst-case and average loss, respectively, apart from the multiplier in the latter case. In order to calculate $\boldsymbol{N}(\mathcal{Y})$, it is necessary that $\tilde{\boldsymbol{F}}^{\mathcal{Y}}\left(\tilde{\boldsymbol{F}}^{\mathcal{Y}}\right)^T$ is regular. In other words, $\tilde{\boldsymbol{F}}^{\mathcal{Y}}$ must not be a tall matrix and must have full row rank (linearly independent rows). A sufficient condition for this requirement is that all implementation errors are non-zero, *i.e.*, $\text{rank}(\boldsymbol{W}_{n^y}) = n_y \Rightarrow \text{rank}(\tilde{\boldsymbol{F}}) = n_y \wedge \text{rank}(\tilde{\boldsymbol{F}}^{\mathcal{Y}}) = n_{\mathcal{Y}}$.

If the B3WC/B3AV algorithm is applied to problem (3.28), then $\boldsymbol{G}_{\boldsymbol{z}}^{\mathcal{Y}}$ and $\tilde{\boldsymbol{F}}^{\mathcal{Y}}$ are substituted by $\boldsymbol{\mathcal{G}}^{\mathcal{Y}}$ and $\boldsymbol{\mathcal{F}}^{\mathcal{Y}}$. In most of the practical relevant cases, $\boldsymbol{\mathcal{F}}^{\mathcal{Y}}$ can be a tall matrix in the downward tree. More precisely, $\text{rank}(\boldsymbol{\mathcal{F}}^{\mathcal{Y}}) \leq n_{\boldsymbol{f}}$ holds always and in the downward tree it is $n_{\mathcal{Y}} \in \{n_u + 1, \ldots, v\}$. As, it may happen that $n_{\boldsymbol{f}} < v$, $\tilde{\boldsymbol{F}}^{\mathcal{Y}}$ may loose full row rank, *i.e.*,

$$\text{rank}(\boldsymbol{\mathcal{F}}^{\mathcal{Y}}) < n_{\mathcal{Y}}. \tag{3.39}$$

Consequently, the B3WC/B3AV algorithm is not applicable in its original form. The code must be modified for the case when (3.39) occurs in the downward tree. If (3.39) is true, (3.37)/(3.38) cannot be evaluated and thus there is no decision criterion for pruning and branching. Downwards pruning can be simply omitted. Downwards branching can be performed based on the row norms of the candidate rows of either $\boldsymbol{\mathcal{G}}$ or $\boldsymbol{\mathcal{F}}$. In this respect, it is important to note that the ratio of the ith row norm of $\boldsymbol{\mathcal{F}}$ and the ith row norm of $\boldsymbol{\mathcal{G}}$ refers to a generalized singular value which is favorable to be small. Thus, the selection of branches are ordered according to the least (or largest) row norms of candidate rows in $\boldsymbol{\mathcal{F}}$ (or $\boldsymbol{\mathcal{G}}$). The computational efficiency of such a modified B3WC/B3AV method is decreased as the downward search is only partial bidirectional. As the B3MSV is not required to be modified, it is favorable over the B3WC/B3AV method if efficiency is more crucial than accuracy.

By the use of the GSVD, the range in which downward pruning is active may be increased from $n_{\mathcal{Y}} \leq \text{rank}(\boldsymbol{\mathcal{F}}^{\mathcal{Y}})$ to $n_{\mathcal{Y}} < \text{rank}(\boldsymbol{\mathcal{F}}^{\mathcal{Y}}) + \text{rank}(\boldsymbol{\mathcal{G}}^{\mathcal{Y}})$. This may be done by substitution of (3.37) and (3.38) by

$$\tilde{\mathcal{L}}_{\text{wc}}(\mathcal{Y}) = \frac{1}{2}\sigma_w^2\left(\left(\boldsymbol{\mathcal{F}}^{\mathcal{Y}}\right)^T, \left(\boldsymbol{\mathcal{G}}^{\mathcal{Y}}\right)^T\right)$$

and

$$\tilde{\mathcal{L}}_{\text{av}}(\mathcal{Y}) = \sum_{i=1}^{w}\sigma_i^2\left(\left(\boldsymbol{\mathcal{F}}^{\mathcal{Y}}\right)^T, \left(\boldsymbol{\mathcal{G}}^{\mathcal{Y}}\right)^T\right).$$

respectively. For $n_{\mathcal{Y}} \leq \text{rank}(\boldsymbol{\mathcal{F}}^{\mathcal{Y}})$, they agree with (3.37) and (3.38). For $n_{\mathcal{Y}} \geq \text{rank}(\boldsymbol{\mathcal{F}}^{\mathcal{Y}}) + \text{rank}(\boldsymbol{\mathcal{G}}^{\mathcal{Y}})$ they are both zero. In between, it must be shown that monotonicity holds.

Conjecture 3.14. If $\text{rank}(\boldsymbol{\mathcal{F}}^{\mathcal{Y}}) \leq n_{\mathcal{Y}} \leq \text{rank}(\boldsymbol{\mathcal{F}}^{\mathcal{Y}}) + \text{rank}(\boldsymbol{\mathcal{G}}^{\mathcal{Y}})$ and $w = \text{rank}(\boldsymbol{\mathcal{G}}^{\mathcal{Y}}) \geq n_{\boldsymbol{u}}$, then $\tilde{\mathcal{L}}_{\star}(\mathcal{Y}\backslash i) \geq \tilde{\mathcal{L}}_{\star}(\mathcal{Y})$ holds where $i \in \mathcal{Y}$ and $\star$ refers to either av or wc.

This conjecture was verified by means of repeated random tests. A general proof is an open problem.

3.D. MI problem solution

The decomposed formulation (3.25) gives rise to the efficient AM methods which deliver approximate solutions to the mixed-integer problems (3.13) and (3.16). Unfortunately, no error bound between the approximate and the actual solution could be derived. In order to get an idea of the appropriability of the AM methods, they need to be numerically compared with more accurate solution methods which are developed in this section. The numerical comparison takes place in Section 3.5.

As pointed out in Section 3.4.3, the continuous problem of finding the linear coefficients for a predefined structure has no analytical solution and thus iterative solution methods need to be applied. The decomposed formulation (3.25) is beneficial for iterative solution techniques due to two effects. First, selecting $\tilde{\boldsymbol{h}}_i = \boldsymbol{e}_{n y_i}$ likely gives good starting values for the iteration as it selects the smallest generalized singular values as can be seen by observation of (3.25). Second, the number of optimization variables can be reduced if $r_i > 0$ for any $i \in \{1, \dots n_{\boldsymbol{u}}\}$ as already discussed in Remark 3.13. It is thus convenient to keep the formulation (3.25) and perform the iterative search in a reduced transformed space.

In order to improve the efficiency of the search of the best set in the superstructure, BAB methods can be applied. BAB methods search the best set in a solution tree where each branch end relates to a full set and each node relates to an incomplete set. A prerequisite for the application of BAB methods is that the objective to be minimized, *i.e.*, L_{wc} or L_{av}, can be replaced by function which can be evaluated also for incomplete sets, which equals the objective at full sets and which is monotonically increasing from the tree root towards the branch ends.

Definition 3.15. Let

$$\mathcal{L}_k = \min_{\{\tilde{\boldsymbol{h}}_1, \dots, \tilde{\boldsymbol{h}}_k\}} \left\| \boldsymbol{\mathcal{F}}_k \, \boldsymbol{\mathcal{X}}_k^{-1} \right\|_\star \tag{3.40}$$

to be the global optimal value of the continuous problem of a particular set of a superstructure and $\tilde{\boldsymbol{h}}_1^*, \dots, \tilde{\boldsymbol{h}}_k^*$ the relating solution. Here,

$$\begin{aligned}
\boldsymbol{\mathcal{F}}_k &= \sum_{i=1}^{k} \tilde{\boldsymbol{U}}_i \, \tilde{\boldsymbol{\Sigma}}_i \, \tilde{\boldsymbol{h}}_i \, \boldsymbol{e}_i^T \\
\boldsymbol{\mathcal{X}}_k &= \bar{\boldsymbol{\mathcal{X}}}_k + \begin{bmatrix} \boldsymbol{0}_{n_u \times k} & \ker\left(\bar{\boldsymbol{\mathcal{X}}}_k^T\right) \end{bmatrix} \\
\bar{\boldsymbol{\mathcal{X}}}_k &= \sum_{i=1}^{k} \boldsymbol{U}_i \, \boldsymbol{\Sigma}_i \, \tilde{\boldsymbol{h}}_i \, \boldsymbol{e}_i^T
\end{aligned}$$

and $\star$ refers to either the induced 2-norm or the Frobenius norm. It is important to stress that $\mathcal{L}_{n_u}^2$ equals either $2\, L_{\mathrm{wc}}$ or $6\,(n y_{\mathrm{u}} + n_d)\, L_{\mathrm{av}}$ depending on the specification of

$\star$. Further,

$$\mathcal{L}_k^- = \left\|\boldsymbol{\mathcal{F}}_{k-1}\boldsymbol{\mathcal{X}}_{k-1}^{-1}\right\|_\star \mid \tilde{\boldsymbol{h}}_i = \tilde{\boldsymbol{h}}_i^* \forall i \in \{1,\ldots,k-1\}$$

is introduced. $\mathcal{L}_k^-$ is almost identical to the solution of (3.40) except for the fact that $\tilde{\boldsymbol{h}}_k$ is a zero vector.

Lemma 3.16. *For $k \geq 2$, the inequality*

$$\mathcal{L}_k^- \leq \mathcal{L}_k \tag{3.41}$$

holds irrespective of the specification of $\star$.

Proof. Performing the Gram-Schmidt decomposition of $\bar{\boldsymbol{\mathcal{X}}}_k$ yields $\bar{\boldsymbol{\mathcal{X}}}_k = \boldsymbol{Q}\boldsymbol{R}$ with the unitary $\boldsymbol{Q}$ and the upper triangular $\boldsymbol{R} \in \mathbb{R}^{n_u \times k}$. As the last $n_{\boldsymbol{u}} - k$ columns of $\boldsymbol{Q}$ equal $\ker\left(\bar{\boldsymbol{\mathcal{X}}}_k\right)$, one can write

$$\boldsymbol{\mathcal{X}}_k = \boldsymbol{Q}\begin{bmatrix} \boldsymbol{R}_k & \\ & \boldsymbol{I}_{n_u-k}\end{bmatrix},$$

where $\boldsymbol{R}_i \in \mathbb{R}^{i\times i}\,\forall 1 \leq i \leq k$ is a principal minor of $\boldsymbol{R}$. Recall that due to the upper triangular form of $\boldsymbol{R}$, the decomposition is preserved if the last column in $\bar{\boldsymbol{\mathcal{X}}}_k$ is deleted with the only difference that the last column in $\boldsymbol{R}$ is deleted as well (see Golub and VanLoan, 1996, Chapter 12.5.2). Thus, if $k \geq 2$ one can simply conclude to

$$\boldsymbol{\mathcal{X}}_{k-1} = \boldsymbol{Q}\begin{bmatrix} \boldsymbol{R}_{k-1} & \\ & \boldsymbol{I}_{n_u+1-k}\end{bmatrix}.$$

As the inverse of $\boldsymbol{R}_k$ can be calculated from backward substitution, the principal minors of $\boldsymbol{R}_k^{-1}$ depends only on the corresponding principal minors of $\boldsymbol{R}_k$. Accordingly, $\boldsymbol{R}_{k-1}^{-1}$ is a principal minor of $\boldsymbol{R}_k^{-1}$ as $\boldsymbol{R}_{k-1}$ is a principal minor of $\boldsymbol{R}_k$. It is then obvious that the first $k-1$ columns in both

$$\begin{aligned}\boldsymbol{A}_k = \boldsymbol{\mathcal{F}}_k\boldsymbol{\mathcal{X}}_k^{-1}\boldsymbol{Q} &= \boldsymbol{\mathcal{F}}_k\begin{bmatrix} \boldsymbol{R}_k^{-1} & \\ & \boldsymbol{I}_{n_u-k}\end{bmatrix}\\ \boldsymbol{A}_{k-1} = \boldsymbol{\mathcal{F}}_{k-1}\boldsymbol{\mathcal{X}}_{k-1}^{-1}\boldsymbol{Q} &= \boldsymbol{\mathcal{F}}_{k-1}\begin{bmatrix} \boldsymbol{R}_{k-1}^{-1} & \\ & \boldsymbol{I}_{n_u+1-k}\end{bmatrix}\end{aligned}$$

are identical. Thus, the matrix $\boldsymbol{A}_{k-1}^T\boldsymbol{A}_{k-1}$ is a principal minor of $\boldsymbol{A}_k^T\boldsymbol{A}_k$ and Cauchy's interlacing inequalities (Hogben, 2007, pp. 8.3-8.4) apply, *i.e.*,

$$\sigma_1\left(\boldsymbol{A}_k\right)\cdots \leq \sigma_j\left(\boldsymbol{A}_k\right) \leq \sigma_j\left(\boldsymbol{A}_{k-1}\right) \leq \sigma_{j+1}\left(\boldsymbol{A}_k\right) \leq \cdots \leq \sigma_k\left(\boldsymbol{A}_k\right).$$

As $\boldsymbol{A}_k$ and $\boldsymbol{A}_{k-1}$ are similarity transformations of $\boldsymbol{\mathcal{F}}_k\boldsymbol{\mathcal{X}}_k^{-1}$ and $\boldsymbol{\mathcal{F}}_{k-1}\boldsymbol{\mathcal{X}}_{k-1}^{-1}$, respectively, which do not affect the singular values, the inequalities can also be written in terms of $\boldsymbol{\mathcal{F}}_k \boldsymbol{\mathcal{X}}_k^{-1}$ and $\boldsymbol{\mathcal{F}}_{k-1} \boldsymbol{\mathcal{X}}_{k-1}^{-1}$. In particular it holds that

$$\sigma_{k-1}\left(\boldsymbol{\mathcal{F}}_{k-1} \boldsymbol{\mathcal{X}}_{k-1}^{-1}\right) \leq \sigma_k\left(\boldsymbol{\mathcal{F}}_k \boldsymbol{\mathcal{X}}_k^{-1}\right)$$

and

$$\sum_{i=1}^{k-1} \sigma_i\left(\boldsymbol{\mathcal{F}}_{k-1} \boldsymbol{\mathcal{X}}_{k-1}^{-1}\right) \leq \sum_{i=1}^{k} \sigma_k\left(\boldsymbol{\mathcal{F}}_k \boldsymbol{\mathcal{X}}_k^{-1}\right)$$

which correspond to $\left\|\boldsymbol{\mathcal{F}}_{k-1} \boldsymbol{\mathcal{X}}_{k-1}^{-1}\right\|_\star \leq \left\|\boldsymbol{\mathcal{F}}_k \boldsymbol{\mathcal{X}}_k^{-1}\right\|_\star$ and (3.41) for either specification of $\star$. □

Theorem 3.17. *The function $\mathcal{L}_k$ as defined in* (3.40) *is monotonically increasing in k.*

Proof. Per definition, $\mathcal{L}_{k-1}$ is global minimum and thus $\mathcal{L}_{k-1} \leq \mathcal{L}_k^-$. By using (3.41), it follows $\mathcal{L}_{k-1} \leq \mathcal{L}_k$. □

The MIWC/MIAV method refers to a unidirectional BAB search in the upward tree[3] where in each node the problem (3.40) is iteratively solved. The branch with the smallest generalized singular value is always preferred. The upper bound is updated if a full set with a lower loss is found. At the beginning the upper bound is set to infinity[4]. Pruning of branches takes place if $\mathcal{L}_k$ exceeds the upper bound. Due to better accuracy, the MIWC/MIAV method is more favorable than the B3WC-AM/B3WC-AV method in case of small system dimensions ($n_{\boldsymbol{u}}$, $n_{\boldsymbol{d}}$, $n_{\boldsymbol{y}}$) when computational load is not limiting. Moreover, the MIWC/MIAV method has the advantage that it is not restricted to the requirement that for every CV the same candidate CVs need to be taken into account.

Remark 3.18. The decomposed formulation (3.25) is additionally valuable for the a priori elimination of sets in the superstructure as it provides a lower bound for $\mathcal{L}_k$, given by either of both

$$\mathcal{L}_k \geq \max_{i \in \{1,\ldots,k\}} \left(\sigma_{in y_i}\right) \tag{3.42a}$$

$$\mathcal{L}_k \geq \sqrt{\sum_{i=1}^{k} \sigma_{in y_i}^2}. \tag{3.42b}$$

Here, $\sigma_{in y_i}$ refers to the smallest generalized singular value of the ith CV. Both bounds follow from the conclusion that the least $\mathcal{L}_k$ corresponds to the ideal case that $\tilde{\boldsymbol{h}}_i = \boldsymbol{e}_{n y_i} \forall i$,

[3] In an upward tree, the root node refers to an empty CV set

[4] It is worth mentioning that in some cases a finite initial upper bound can be calculated from (3.25).

$\boldsymbol{U}_{i\mathcal{Y}_i} \perp \boldsymbol{U}_{j\mathcal{Y}_j} \forall i \neq j$ and $\tilde{\boldsymbol{U}}_{i\mathcal{Y}_i} \perp \tilde{\boldsymbol{U}}_{j\mathcal{Y}_j} \forall i \neq j$ which yields a diagonal matrix

$$\boldsymbol{\mathcal{F}}_k \boldsymbol{\mathcal{X}}_k^{-1} = \sum_{i=1}^{k} \sigma_{in_{\mathcal{Y}_i}} \boldsymbol{e}_i \boldsymbol{e}_i^T$$

from which (3.42a) and (3.42b) can be trivially deduced. Sets whose lower bound exceeds the upper bound of the solution of the mixed-integer problem can be omitted.

Bibliography

MATLAB documentation, 2008b. URL http://www.mathworks.com.

V. Alstad and S. Skogestad. The null space method for selecting optimal measurement combinations as controlled variables. *Industrial and engineering chemistry research*, 46 (3):846–853, 2007.

V. Alstad, S. Skogestad, and E. S. Hori. Optimal measurement combinations as controlled variables. *Journal of process control*, 19(1):138–148, 2009.

Y. Cao. Bidirectional branch and bound minimum singular value solver: V2, 2007/11/11. URL http://www.mathworks.com/matlabcentral/fileexchange/17480.

Y. Cao. Bidirectional branch and bound solvers for worst case loss minimization, 2009/01/09. URL http://www.mathworks.com/matlabcentral/fileexchange/22632.

Y. Cao. Bidirectional branch and bound for average loss minimization, 2009/11/17. URL http://www.mathworks.com/matlabcentral/fileexchange/25870.

Y. Cao and V. Kariwala. Bidirectional branch and bound for controlled variable selection: Part I. Principles and minimum singular value criterion. *Computers and chemical engineering*, 32(10):2306–2319, 2008.

S. Engell, T. Scharf, and M. Völker. A methodology for control structure selection based on rigorous process models. IFAC world congress, Prague, Czech Republic, 2005/07/04-08.

G. H. Golub and C. F. VanLoan. *Matrix computations*. Johns Hopkins studies in the mathematical sciences. Johns Hopkins Univ. Press, Baltimore, Maryland, 3rd ed. edition, 1996. ISBN 0801854148.

I. J. Halvorsen, S. Skogestad, J. C. Marud, and V. Alstad. Optimal selection of controlled variables. *Industrial and engineering chemistry research*, 42:3273–3284, 2003.

S. Heldt. On a new approach for self-optimizing control structure design. In S. Engell and Y. Arkun, editors, *ADCHEM 2009: Preprints of IFAC symposium on advanced control of chemical processes: July 12-15, 2009, Koç University, Istanbul, Turkey*, volume 2, pages 807–812. 2009.

S. Heldt. Dealing with structural constraints in self-optimizing control engineering. *Journal of process control*, 20(9):1049–1058, 2010a.

L. Hogben. *Handbook of linear algebra*. Discrete mathematics and its applications. Chapman & Hall/CRC, Boca Raton, Florida, 2007. ISBN 1584885106.

R. A. Horn and C. R. Johnson. *Matrix analysis*. Cambridge University Press, Cambridge, UK, 1990. ISBN 0521386322.

J. E. P. Jäschke and S. Skogestad. Optimally invariant variable combinations for nonlinear systems. In S. Engell and Y. Arkun, editors, *ADCHEM 2009: Preprints of IFAC symposium on advanced control of chemical processes: July 12-15, 2009, Koç University, Istanbul, Turkey*, volume 2, pages 551–556. 2009.

J. E. P. Jäschke, H. Smedsrud, S. Skogestad, and H. Manum. Optimal operation of a waste incineration plant for district heating. American control conference, Hyatt Regency Riverfront, St. Louis, Missouri, 2009/06/10-12.

V. Kariwala. Optimal measurement combination for local self-optimizing control. *Industrial and engineering chemistry research*, 46(46):3629–3634, 2007.

V. Kariwala and Y. Cao. Bidirectional branch and bound for controlled variable selection: Part II. Exact local method for self-optimizing control. *Computers and chemical engineering*, 2009.

V. Kariwala and Y. Cao. Bidirectional branch and bound for controlled variable selection: Part III. Local average loss minimization. *IEEE transactions on industrial informatics*, 2010a.

V. Kariwala and S. Skogestad. Branch and bound methods for control structure design. Symposium on Process Systems Engineering/European Symposium on Computer Aided Process Engineering, Garmisch-Partenkirchen, Germany, 2006/07/09-13.

V. Kariwala, Y. Cao, and S. Janardhanan. Local self-optimizing control with average loss minimization. *Industrial and engineering chemistry research*, 47(4):1150–1158, 2008.

T. Larsson, K. Hestetun, E. Hovland, and S. Skogestad. Self-optimizing control of a large-scale plant: The Tennessee Eastman process. *Industrial and engineering chemistry research*, 40(22):4889–4901, 2001.

F. A. Michelsen, B. F. Lund, and I. J. Halvorsen. Selection of optimal, controlled variables for the TEALARC LNG process. *Industrial and engineering chemistry*, 49(18):8624–8632, 2010.

M. Morari, Y. Arkun, and G. Stephanopoulos. Studies in the synthesis of control structures for chemical processes: Part I: Formulation of the problem. Process decomposition and the classification of the control tasks. Analysis of the optimizing control structures. *American institute of chemical engineers journal*, 26(2):220–232, 1980.

C. C. Paige and M. A. Saunders. Towards a generalized singular value decomposition. *SIAM journal on numerical analysis*, 18(3):398–405, 1981.

S. Skogestad. Plantwide control: The search for the self-optimizing control structure. *Journal of process control*, 10(5):487–507, 2000.

S. Skogestad and I. Postlethwaite. *Multivariable feedback control: Analysis and design*. Wiley, Chichester, UK, 1996. ISBN 0471942774.

R. Yelchuru and S. Skogestad. MIQP formulation for optimal controlled variable selection in self-optimizing control. The international symposium on design, operation and control of chemical processes, Singapore, Republic of, 2010/07/25-28.

4. Practical application of CSD methods

This chapter deals with the practical application of the newly developed CSD methods to industrial (nonlinear) plant models. In Section 4.1, a brief survey of recent work in this field is given. Section 4.2 deals with the general practical issues related to self-optimizing CSD. In Section 4.3, self-optimizing CSD for an evaporation process is performed for illustration purposes. Note that this has been previously considered by other authors (Cao, 2004/01/11-14; Agustriyanto and Zhang, 2006/08/30-09/01; Kariwala *et al.*, 2008). General thoughts regarding control of LNG liquefaction processes are carried out in Section 4.4. The results of the application of the newly developed CSD methods to the SMR cycle and to Linde's proprietary LNG liquefaction processes LIMUM® cycle and MFC® process are presented in Section 4.5, 4.6 and 4.7, respectively. Conclusions regarding self-optimizing CSD for LNG liquefaction processes are drawn in Section 4.8.

4.1. Related work

During the last decade, there has been a considerable amount of works published which are in some way or another dedicated to the practical application of self-optimizing CSD. Here, a brief survey of selected works is presented.

By economic considerations, Larsson *et al.* (2001) addresses the problem of finding a self-optimizing control structure for the benchmark process of Tennessee Eastman Company. Their CSD procedure was based on heuristical rules and process insight. The resulting control structure was verified by the use of dynamical simulation. Similarly, Engell *et al.* (2005/07/04-08) proposed the successive disregarding of PVs as candidate CVs based on a specially derived measure. They tested their results on a reactive distillation column example. The resulting control structure was checked via dynamic simulation studies of the linearized process model. Hori *et al.* (2006/09/04-06) and Hori and Skogestad (2008) designed self-optimizing control structures based on the MSV rule for two-product distillation columns. They considered indirect composition control of binary and multiphase mixtures, evaluated control structures with combinations of maximum two to four PVs and concluded heuristical design rules. In their first paper, dynamical simulation served as a verification of the results. Based on the same process model, Kariwala and Cao (2010a) derived the best column control structures by the use of the efficient PB3AV method where PV set sizes from minimum to maximum were imposed. Self-optimizing CSD for the evaporation process treated in Section 4.3 was the subject of the work of Cao (2004/01/11-14); Agustriyanto and Zhang (2006/08/30-09/01) and Kariwala *et al.* (2008). The CSD of the former authors is based on solving a constrained optimization problem, while the latter use average loss minimization. Validation took place by dynamical simulation (former two papers) and nonlinear steady-state evaluation (latter paper). Lersbamrungsuk (2008) investigated the optimal operation of simplified heat exchanger networks with regulatory control (nonlinear optimization problem) where switching between operating regions is essential. A control structure based on split-range control was derived for a case study and practicability was proven by dynamic simulation studies. Self-optimizing control of simple heat pump and refrigeration cycles as well as an LNG liquefaction processes (PRICO) has been investigated by Jensen (2005/05/29-06/01); Jensen and Skogestad (2007a, 2009a). Alternative control structures were judged by both self-optimizing control measures (MSV) and non-linear parameter studies. Baldea *et al.* (2008) considered CSD for a reactor-separator process and took both, steady-state economics and dynamical behavior, into account. Their result was successfully tested via dynamic simulation studies. Oldenburg *et al.* (2008/06/01-04) presented a self-optimizing control structure selection method based on simple and efficient re-use of steady-state simulation data. By retrospective view on practically operating control examples, Downs and Skogestad (2009) showed that self-optimizing control has been successfully applied in industry. Almost optimal steady-state operation of nonlinear process examples is the subject of the works of Jäschke *et al.* (2009/06/10-12) and Jäschke and Skogestad (2009). Based on an analytical process model, they proposed a new framework for finding CVs (referred to as invariants) as nonlinear combinations of PVs and switching conditions between operating regions. They considered a waste incineration plant (first paper) and a continuous stirred-tank reactor with two parallel reactions (second paper). For verification of their results, dynamical simulation (first paper) and steady-state parameter studies (second paper) were carried out. Michelsen *et al.* (2010) applied self-optimizing control structure design algorithms to a proprietary SMR cycle and verified their results with nonlinear tests.

4.2. CSD practice

In this section, the workflow for the design of self-optimizing control structures for industrial plants is described (Section 4.2.1) and the sourcing of necessary information is pointed out (Section 4.2.2).

4.2.1. Workflow

Plantwide control design procedures have been proposed by Skogestad (2004a) and Downs and Skogestad (2009), among others. In this section, the general practice for designing self-optimizing control structures is discussed from the point of view of a plant constructor which is confronted with the difficulty to deliver them in an early phase of plants' life cycles. The corresponding workflow is represented in Figure 4.1. The basis of all is the design process flowsheet/model of the plant, representing the most common operating case. It needs to be transformed into a rating flowsheet/model which includes equipment dimensions, performance maps for rotating equipment and models for phenomenologies such as heat/mass transfer and frictional pressure drop. Take into account that it is often convenient to reduce the number of material species to a minimum to enhance the computational performance of the rating flowsheet.

Self-optimizing control depends on an objective to be optimized. If there are more MVs present than product/operational specifications, then the surplus MVs can be used for optimizing an objective J such as energy consumption and feed throughput. If less MVs are present than product/operational specifications, then the achievement of all targets is infeasible but all MVs can be used for minimizing the deviation from the target solution, for instance in the sense of a weighted least-squares sum.

Remark 4.1. Suppose a design flowsheet which is already optimal with respect to a certain objective was transformed into a rating flowsheet. It is sometimes a matter of discussion whether the resulting rating flowsheet is still in its optimal state. This is generally not the case, as on the one hand model equations have changed and on the other hand optimization constraints may have become obsolete. For instance for optimization of design models, minimal temperature differences between the heat exchanging streams are imposed as constraints. These constraints are generally dropped for optimization of rating models as phenomenological model equations for heat transfer introduce a natural barrier for the violation of a minimum temperature difference.

For self-optimizing control, it is necessary that the cost function is convex and that an unconstrained optimum is achieved, *i.e.*, it holds that $J_{\boldsymbol{uu}} \succ \mathbf{0}$ and $J_{\boldsymbol{u}} = \mathbf{0}$, respectively. If the optimization of the rating flowsheet reveals that there are active constraints, then the number of unconstrained MVs has to be reduced until an unconstrained optimum

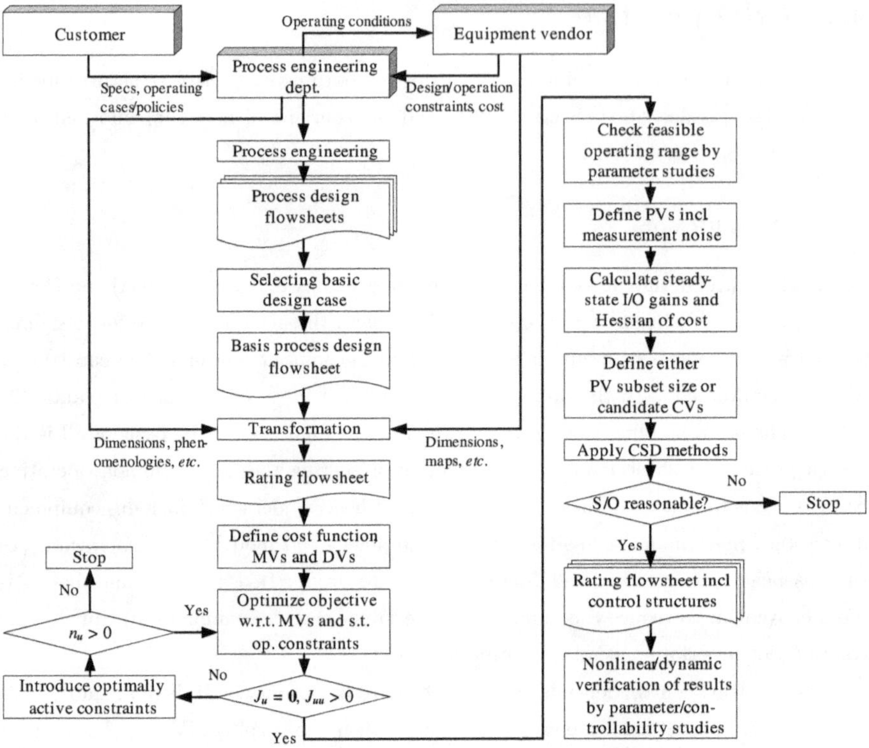

Figure 4.1.: Workflow for self-optimizing CSD

can be found. Examples for active operational constraints are the compressor surge line or a minimum superheating (margin from dew point) at compressor inlet. It is worth mentioning that it is irrelevant for self-optimizing CSD which particular MVs are used for ensuring the active constraints and which not. The ultimate classification comes into play not until control configuration is considered. If active constraints change between certain operability regions, then for each region individual self-optimizing control structures are required (Lersbamrungsuk, 2008; Jäschke and Skogestad, 2009; Jäschke *et al.*, 2009/06/10-12). It is important to stress that due to the need of doing parameter studies, the identification of optimally active constraints over the complete operability range can be time-consuming, especially for large numbers of MVs and DVs.

An estimate of the (feasible) operating region can be obtained by doing parameter studies of the MVs and DVs in each direction until infeasibility occurs. Infeasibility is an attribute of nonlinear steady-state models and refers to a situation where a set of equations has no (real) solution. For instance, it may happen that the setpoint of a controller cannot be reached. In reality and dynamic models this results in setpoint deviations, *etc.*

The feasibility region and the expected variances of the DVs give clues for the diagonal scaling matrix $\boldsymbol{W}_d$. Further, it is necessary to determine the implementation errors, present in the diagonal scaling matrix $\boldsymbol{W}_{n^y}$. The I/O gains $\boldsymbol{G}^y$ and $\boldsymbol{G}_d^y$ and the Hessians of the objective with respect to the MVs and DVs can be calculated as pointed out in Section 4.2.2.

With the system information J_d, J_{uu}, J_{ud}, J_{dd} and $\boldsymbol{W}_d$ it can be checked whether optimization of operation is anyway reasonable. The measures introduced by Engell (2007) and presented in Section 4.4.2 give advice for this decision. Further it can be judged whether self-optimizing control is sufficient in terms of close to optimality or whether it may be beneficial to enhance it by setpoint optimization techniques such as RTO.

The system information J_{uu}, J_{ud}, $\boldsymbol{G}^y$, $\boldsymbol{G}_d^y$, $\boldsymbol{W}_d$ and $\boldsymbol{W}_{n^y}$ is sufficient for the application of self-optimizing CSD methods. If PV selection control structures are desired, the application of the B3WC/B3AV method is recommended (Kariwala and Cao, 2009, 2010a). If column or common-sized control structures are desired, it must be decided on the maximum number of combined PVs ny. Column control structures can be obtained by application of the PB3WC/PB3AV method (Kariwala and Cao, 2009, 2010a). If common- and individually-sized control structures are aimed, the B3WC-AM/B3AV-AM method needs to be applied. This method requires the specification of structural candidate CVs. This can either be done by selecting them from process insight or by specifying the maximum number of combined PVs ny and further structural constraints such as whether mixed-unit combinations are allowed *etc.*

The resulting self-optimizing control structures need to be discriminated with respect to nonlinear model behavior. Control structures which on the one hand provide a sufficiently large feasibility region and on the other hand show small deviation from the ideal cost function are preferred. Nonlinear studies as a means can be time-consuming for large-scale models and have apart from this work only been publicly considered by Jensen and Skogestad (2007a, 2009b). A further possibility for the judgment of nonlinear behavior is the use of Monte Carlo simulations as applied by Kariwala *et al.* (2008). However, this is not considered in this work.

The resulting control structures can be judged optionally by dynamical considerations. This is outside the scope of the workflow indicated in Figure 4.1 as it requires a dynamical rating flowsheet/model, but nevertheless subject of this work. Chapter 5 is dedicated to the verification of control structures for LNG liquefaction processes by means of dynamical measures.

4.2.2. Data acquisition

As pointed out above, the locally exact identification methods for self-optimizing control structures are based on a linear I/O model (3.2) and a quadratic cost function (3.3) of the plant. This section deals with the acquisition of the required data. Steady state models of industrial plants follow from first principles and are usually obtained by utilization of flowsheeting tools such as OPTISIM®. As presented in Eich-Soellner *et al.* (1997), these models are represented by the generic form

$$\mathbf{0} = \boldsymbol{f}\left(\boldsymbol{x}, \boldsymbol{p}, \operatorname{sgn} \boldsymbol{q}, \boldsymbol{v}\right) \tag{4.1a}$$

$$\text{s.t. } \boldsymbol{q} = \boldsymbol{q}\left(\boldsymbol{x}, \boldsymbol{p}\right), \tag{4.1b}$$

where $\boldsymbol{f} \in \mathbb{R}^{n_{\boldsymbol{x}}}$ refers to a nonlinear function with all time derivatives set to zero. The state vector $\boldsymbol{x} \in \mathbb{R}^{n_{\boldsymbol{x}}}$ collects state variables such as concentrations, material and energy flow rates, temperatures, pressures. The vector $\boldsymbol{p} \in \mathbb{R}^{n_{\boldsymbol{p}}}$ summarizes invariant model parameters, *e.g.*, heat exchanger areas. The so-called switching functions $\boldsymbol{q} \in \mathbb{R}^{n_{\boldsymbol{q}}}$ (4.1b) are used to switch among different states of the system described by different subsets of equations. Typical reasons are saturation (*e.g.*, in control devices), the appearance or disappearance of phases, piecewise-continuous physical-property correlations, *etc.* The vector $\boldsymbol{v} \in \mathbb{R}^{n_{\boldsymbol{v}}}$ refers to the degrees of freedom of the plant. It can be divided into two classes of variables, *i.e.*, $\boldsymbol{v} = \begin{bmatrix} \boldsymbol{u}^T & \boldsymbol{d}^T \end{bmatrix}^T$ and $n_{\boldsymbol{v}} = n_{\boldsymbol{u}} + n_{\boldsymbol{d}}$. Here, $\boldsymbol{u}$ indicates the independent input variables which may serve as MVs for control purposes and $\boldsymbol{d}$ are the dependent input variables usually referred to as DVs.

Note that all equations in (4.1a) in which an element of $\boldsymbol{v}$ is involved can be summarized as

$$\mathbf{0} = \boldsymbol{P}_{\text{in}}\, \boldsymbol{f} = \boldsymbol{x}_{\boldsymbol{v}} - \boldsymbol{v}, \tag{4.2}$$

where $\boldsymbol{x}_{\boldsymbol{v}} \subseteq \boldsymbol{x}$ and $\boldsymbol{P}_{\text{in}}$ is a binary matrix with one one per row. The model outputs are a subset of state variables, *i.e.*, $\boldsymbol{y} \subseteq \boldsymbol{x}$, and are thus obtained by

$$\boldsymbol{y} = \boldsymbol{P}_{\text{out}}\, \boldsymbol{x}, \tag{4.3}$$

where $\boldsymbol{P}_{\text{out}}$ is a binary matrix with one one per column. The total derivative of (4.1a) at the operating point $\{\boldsymbol{x}_0, \boldsymbol{v}_0\}$ reads

$$\mathrm{d}\boldsymbol{f} = \left.\frac{\partial \boldsymbol{f}}{\partial \boldsymbol{x}^T}\right|_0 \mathrm{d}\boldsymbol{x} + \left.\frac{\partial \boldsymbol{f}}{\partial \boldsymbol{v}^T}\right|_0 \mathrm{d}\boldsymbol{v}.$$

By requiring $\mathrm{d}\boldsymbol{f} = \mathbf{0}$, it follows that

$$\frac{\mathrm{d}\boldsymbol{x}}{\mathrm{d}\boldsymbol{v}^T} = - \left.\frac{\partial \boldsymbol{f}}{\partial \boldsymbol{x}^T}\right|_0^{-1} \left.\frac{\partial \boldsymbol{f}}{\partial \boldsymbol{v}^T}\right|_0 = \left.\frac{\partial \boldsymbol{f}}{\partial \boldsymbol{x}^T}\right|_0^{-1} \boldsymbol{P}_{\mathrm{in}}^T,$$

where the last step follows from observation of (4.2). According to (4.3),

$$\frac{\mathrm{d}\boldsymbol{y}}{\mathrm{d}\boldsymbol{v}^T} = \boldsymbol{P}_{\mathrm{out}} \frac{\mathrm{d}\boldsymbol{x}}{\mathrm{d}\boldsymbol{v}^T} = \boldsymbol{P}_{\mathrm{out}} \left.\frac{\partial \boldsymbol{f}}{\partial \boldsymbol{x}^T}\right|_0^{-1} \boldsymbol{P}_{\mathrm{in}}^T.$$

This result relates to the linear I/O gains as present in (3.2), *i.e.*, $\mathrm{d}\boldsymbol{y}/\mathrm{d}\boldsymbol{v}^T = \begin{bmatrix} \boldsymbol{G}^y & \boldsymbol{G}_d^y \end{bmatrix}$. If an equation-oriented process simulator such as OPTISIM® is used, the Jacobian $\partial \boldsymbol{f}/\partial \boldsymbol{x}|_0$ is directly available and so are $\boldsymbol{G}^y$ and $\boldsymbol{G}_d^y$. Otherwise, the I/O gains must be estimated by finite differences.

The second necessary information besides the I/O gains are the evaluations of the second derivatives of the cost/profit function at the operating point, *i.e.*, $J_{\boldsymbol{uu}}$, $J_{\boldsymbol{ud}} = J_{\boldsymbol{du}}^T$ and $J_{\boldsymbol{dd}}$. As the second derivatives are usually not directly available in a process simulator, they need to be estimated by finite differences. If only functional values are taken into account, *e.g.*, by the use of a sequential simulator, then one element of the Hessian can be calculated by the forward-difference formula given by Dennis and Schnabel (1983, p. 80)

$$\frac{\partial^2 J}{\partial v_i \partial v_j} = \frac{1}{h^2} \left(J(\boldsymbol{v}) - J(\boldsymbol{v} + \boldsymbol{e}_i h) - J(\boldsymbol{v} + \boldsymbol{e}_j h) + J(\boldsymbol{v} + \boldsymbol{e}_i h + \boldsymbol{e}_j h) \right) + \mathcal{O}(h),$$

which requires $n_{\boldsymbol{u}} + n_{\boldsymbol{d}} + \frac{1}{2}(n_{\boldsymbol{u}} + n_{\boldsymbol{d}})^2$ functional evaluations in order to obtain the full (symmetric) Hessian. If an equation-oriented simulator is used, only $n_{\boldsymbol{u}} + n_{\boldsymbol{d}}$ gradient calls are required instead. One element of the Hessian is then calculated by the forward-difference formula

$$\frac{\partial^2 J}{\partial v_i \partial v_j} = \frac{1}{2h} \left(\left.\frac{\partial J}{\partial v_j}\right|_{\boldsymbol{v}+\boldsymbol{e}_i h} - \left.\frac{\partial J}{\partial v_j}\right|_{\boldsymbol{v}} \right) + \frac{1}{2h} \left(\left.\frac{\partial J}{\partial v_i}\right|_{\boldsymbol{v}+\boldsymbol{e}_j h} - \left.\frac{\partial J}{\partial v_i}\right|_{\boldsymbol{v}} \right) + \mathcal{O}(h).$$

given by Dennis and Schnabel (1983, p. 80). If function/gradient evaluations are rather inexpensive and higher accuracy is desired, central difference approximations with an error bound of $\mathcal{O}(h^2)$ may be applied as presented in the textbook of Abramowitz and Stegun (1970, p. 884).

An interesting question is how large should the finite difference interval be chosen? This is related to finding the best trade-off between relative truncation error (tends to increase with interval size) and the relative condition error (generally decreasing with

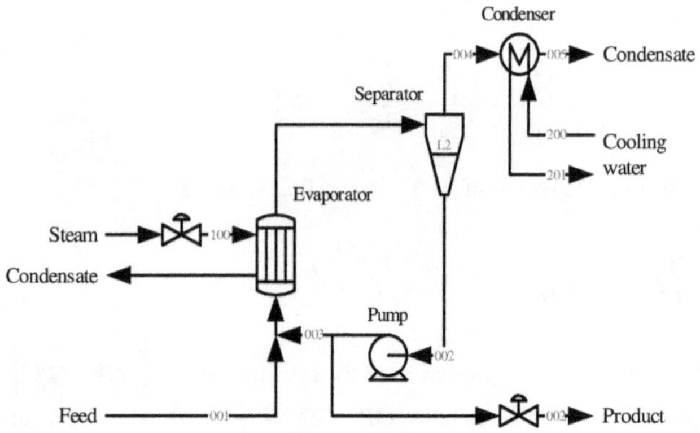

Figure 4.2.: Evaporation process scheme

interval size). Gill *et al.* (1981) proposed a solution approach for this problem. However, in order to account for higher order nonlinearities over the complete feasible region, it is suggested selecting the intervals as large as possible. As the curvature of the cost function may be asymmetric, it is recommended to evaluate it in both directions and select the direction with the larger bend. Alternatively, both directions may be taken into account by minimizing the least-squares sum of the differences of the evaluated functional/gradient values and their (Taylor series) prediction.

4.3. Evaporator case study

In this section, the newly proposed CSD methods are applied to the evaporation process presented in Figure 4.2. This forced-circulation evaporation was originally treated by Newell and Lee (1989) and has been investigated subsequently by Heath *et al.* (2000); Cao (2004/01/11-14) and Kariwala *et al.* (2008), among others. The purpose of the process is the concentration of dilute liquor from the feed to the product stream by evaporation and separation of the solvent. The analytic model equations (4.8a) through (4.8l) are presented in Appendix 4.B. The process model has three state variables, the level L_2, the composition X_{002} and the pressure p_{002} with eight degrees of freedom. Table 4.1 lists the important stream properties, their value at the nominal operating point and their classification into MVs, DVs and PVs. Three out of five MVs indicated by † are used to keep the three PVs L_2, X_{002} and p_{100} at their setpoints. Note that the level L_2 in the separator has no steady-state effect but needs to be controlled for stabilization. The other two controlled PVs need to be kept at their constraints in order to achieve optimality over the given disturbance region. Generality is not lost by this particular selection of the unconstrained MVs. In Table 4.1, the (embraced) expected variations of the DVs

Variable	Description	Nominal value	Classification
F_{001}	Feed flow rate	9.469kg/min	MV, PV (±2 %)
F_{002}	Product flow rate	1.334 kg/min	MV†, PV (±2 %)
F_{003}	Circulating flow rate	24.721 kg/min	MV†, PV (±2 %)
F_{004}	Vapor flow rate	8.135 kg/min	
F_{005}	Condensate flow rate	8.135 kg/min	PV (±2 %)
X_{001}	Feed composition	5.00 %	DV (±5 %)
X_{002}	Product composition	35.50 %	
T_{001}	Feed temperature	40.0 °C	DV (±20 %)
T_{002}	Product temperature	88.4 °C	PV (±1 °C)
T_{003}	Vapor temperature	81.066 °C	PV (±1 °C)
p_{002}	Operating pressure	51.412 kPa	PV (±2.5 %)
F_{100}	Steam flow rate	9.434 kg/min	PV (±2 %)
T_{100}	Steam temperature	151.52 °C	
p_{100}	Steam pressure	400.0 kPa	MV†
Q_{100}	Heat duty	345.292 kW	
F_{200}	Cooling water flow rate	217.738 kg/min	MV, PV (±2 %)
T_{200}	Water inlet temperature	25.0 °C	DV (±20 %)
T_{201}	Water outlet temperature	45.55 °C	PV (±1 °C)
Q_{200}	Condenser duty	313.21 kW	
J	Cost	-582.233 $/h	

Table 4.1.: Process variables of the evaporation process

and measurement errors of the PVs are given in % from their nominal value except for temperature measurement errors indicated on an absolute scale.

The economic objective is to maximize the operational profit given by the unit equation

$$J_{\mathrm{p}} = (4800\, F_{002} - 0.2\, F_{001} - 600\, F_{100} - 0.6\, F_{200} - 1.009\, (F_{002} + F_{003})) \left[\frac{\$/\mathrm{h}}{\mathrm{kg/min}}\right].$$

Here, the first term is the income due to product sale. The second term is the raw material cost. The latter three terms refer to the operational cost of the steam, the water and the pumping, respectively. For convenience, the optimization problem is stated as a minimization of the negative profit $J = -J_{\mathrm{p}}$.

The operational constraints (4.9a) through (4.9f) presented in Appendix 4.B are imposed (for details see Kariwala *et al.*, 2008). The model equations were implemented in a modeling environment (ME) of the Linde in-house simulator OPTISIM®. The model was optimized with respect to the DVs' nominal values given in Table 4.1 and the operational constraints (4.9a) through (4.9f). This led to the operating point presented in Table 4.1. As the ME provides first derivatives by automatic differentiation, the linear I/O gains $G_{\boldsymbol{u}}^{\boldsymbol{y}}$ and $G_{\boldsymbol{u}}^{\boldsymbol{y}}$ at the operating conditions were directly available. Second derivatives $J_{\boldsymbol{uu}}$ and

$n_{\mathcal{Y}}$	Best PV set	L_{av} (in \$/h)	L_{wc} (in \$/h)
2	F_{003}, F_{200}	3.8079	56.7126
3	F_{002}, F_{100}, F_{200}	0.6533	11.6643
4	F_{002}, T_{201}, F_{003}, F_{200}	0.4545	9.4516
...			
10	All PVs	0.1941	7.5015

Table 4.2.: Worst-case/average loss of best column control structures for the evaporation process

J_{ud} were estimated by finite difference approximation and the use of the NAG® routine E04XAF. The numerical results of the I/O gains and the Hessian are in agreement with those of Kariwala *et al.* (2008).

Control structures for the evaporation process have been identified using the methods presented in Sections 3.3 and 3.4. The calculations were conducted in MATLAB® R2008b using a Windows XP® SP2 desktop with Intel® Core™ Duo CPU E8400 (3.0 GHz, 3.5 GB RAM).

At first, column control structures were identified by average loss minimization using the PB3AV method. Some results are given in Table 4.2. They reproduce the results by Kariwala *et al.* (2008) with a deviation of less than 0.6 %. Both, the minimum worst-case and average loss of the best structure decrease with the PV subset size and approach a lower bound (at $n_{\mathcal{Y}} = 10$) asymptotically. According to Corollary 3.8, the case $n_{\mathcal{Y}} = n_{\boldsymbol{u}} = 2$ indicates as well the best PV selection structure.

Next, common-sized control structure with $n_{\mathcal{Y}}$ PVs were sought. According to Lemma 3.7, for each $n_{\mathcal{Y}}$-sized column control structure there exists a common-sized control structure of size $n_s = (n_{\mathcal{Y}} - n_{\boldsymbol{u}} + 1)$ which can be obtained via a simple linear transformation of the former (PB3WC-LT method). Thus, the results in Table 4.2 indicate also possible common-sized control structures of PV subset size n_s from one to nine. For instance, Table 4.3 shows the transformation of the best column control structure with set size three, indicated by COL3, into COM2 where only combinations of two PVs per CV occur. Despite its small PV subset size, COM2 achieves a considerably small worst-case/average loss. Note that control structures obtained by this approach are generally not the best among all control structures satisfying the particular structural constraint of sets with common set size two. In order to find a control structure with lower average loss, the MIAV method was applied. The best solution found among the $\left(C_2^{10}\right)\left(C_2^{10}-1\right)/2 = 990$ alternatives is indicated as COM2† in Table 4.3. Due to the BAB algorithm only 103 problems with an average of 0.07 s expense per problem needed to be solved leading to a total computation time of 8.3 s. Using the B3WC-AM method, the computational efficiency could

Name	Linear control structure	L_{av} (in \$/h)	L_{wc} (in \$/h)
COL3	$\left\{\begin{array}{l} -0.99\,F_{002} + 0.15\,F_{100} + 0.0\,F_{200}, \\ -0.99\,F_{002} - 0.12\,F_{100} + 0.01\,F_{200} \end{array}\right\}$	0.6533	11.6643
COM2	$\{-6.27\,F_{002} + F_{100}, -143.08\,F_{002} + F_{200}\}$	0.6533	11.6643
COM2†	$\{-6.27\,F_{002} + F_{100}, F_{200} - 23.3\,F_{001}\}$	0.5673	11.8034
COM2‡	$\{F_{100} - 0.045\,F_{200}, F_{100} - 6.28\,F_{002}\}$	0.6537	11.666
IND1F	$\{F_{003}, 0.36\,T_{002} + 0.33\,T_{003} + 0.87\,T_{201}\}$	2.9704	57.2328
COL3T	$\left\{\begin{array}{l} 0.59\,T_{002} + 0.53\,T_{003} - 0.61\,T_{201}, \\ 0.02\,T_{002} + 0.01\,T_{003} + 1.0\,T_{201} \end{array}\right\}$	3.6573	57.8693

Table 4.3.: CSD results for the evaporation process

be reduced resulting in a total computation time of 0.15 s at 19 problem evaluations. The related solution is indicated by COM2‡. Due to disregarding of solution space, the solution COM2‡ is more inaccurate and shows larger average loss than both COM2 and COM2†.

Suppose that due to cost issues, only one flow meter can be afforded. As temperature and pressure indicators are rather cheap, their numbers are not limited by cost considerations. Suppose further that mixed unit combinations are disregarded. In this situation, the task is to find the two best CVs out of $C_1^6 + 2$ candidates, *i.e.*, one out of six flows, one pressure and one temperature set. The best control structure indicated as IND1F in Table 4.3 was found by applying the MIAV method. It shows slightly better loss than COL3T which is the column control structure where all temperatures are used.

4.4. General considerations

This section addresses general considerations regarding optimal operation of LNG liquefaction processes. In Section 4.4.1, practical operability issues and their effects on optimal control are discussed. In Section 4.4.2, decision measures introduced by Engell *et al.* (2005/07/04-08) are recalled. Regarding a particular process, these measures are valuable for deciding whether self-optimizing control is sufficient for achieving profit/cost targets or whether a combination with online optimization techniques is reasonable.

4.4.1. LNG liquefaction processes

Objectives and constraints of the operation of LNG liquefaction processes and their implications on self-optimizing control are presented next.

1. The product conditions, *i.e.*, the LNG product pressure and temperature, must be controlled fast and tight. The production rate can be controlled much slower

as the LNG is not directly delivered but stored in large tanks from where it is discontinuously offloaded.

2. The operation range of compressors is constrained by the surge and the choke limit comprehensively described in the textbook of Lüdtke (2004, pp. 140-156). Off-design operation near these limits may cause machinery damage and plant trip, both increasing the plant downtime which is crucial for life cycle cost. Protection mechanisms such as anti-surge control (McMillan, 1983) help to avoid abnormal compressor operation. Nevertheless, some operators consider it more comfortable and safer to leave the compressor regulation unaffected and therefore take the respective controllers into manual operation mode (Mandler and Brochu, Nov 1997). It is therefore a matter of discussion whether or not taking compressor regulation into account for CSD. An advantage of leaving for instance the compressor speed unaffected is the omission of expensive variable speed drivers. It is important to stress that despite constant speed policy, the plant can be operated in turndown operation by opening the anti-surge bypasses.

3. Mandler and Brochu (Nov 1997) and Mandler *et al.* (May 1998) state that it is desirable to control the LNG production rate independently from the LNG temperature. They developed control strategies in which the LNG temperature is controlled by other MVs than LNG flow rate, *e.g.*, compressor speed. Note that this idea is controversial as the operating staff usually follows the traditional concept of adjusting the LNG flow rate in order to fulfill the LNG temperature. This is mainly due to the arguments in points (1) and (2).

4. Dosing of make-up streams (and venting of mixed refrigerant) which regulate the mixed refrigerant composition are not considered available MVs for optimization purposes as they state a rather costly regulation method.

5. In order to decouple the operation of an LNG liquefaction process from ambient temperature variations, it is common practice to keep the temperature of the mixed refrigerant after the ambient coolers fixed, *e.g.*, via variable speed fans attached to aircoolers. If not particularly indicated, omission of temperature regulation is assumed throughout the thesis. This is due to the fact that in most cases a smaller ambient temperature increases the liquefaction capacity of a process.

6. If heavy hydrocarbons (HHCs) are present in the natural gas, they need to be removed in order avoid freezing out in the cold part of the plant and in order to fulfill product specifications of the LNG. Removal of HHCs can be performed within the LNG liquefaction process, *i.e.*, in a vapor-liquid separator behind the

subcooler provided that a subcritical natural gas is present. It is shown below that, from an operational point of view, separation of HHCs within the cycle is unfavorable as one degree of freedom gets lost for optimization. In order to point out the maximum optimization benefit, HHC separation is disregarded in all example processes considered in this work.

7. It is not the objective of this work to design control structures which cover optimal operation for all possible disturbances and operation scenarios. This would require the use of online optimization techniques such as RTO/LMPC. That these are not considered here is due to various reasons discussed in Section 1.1. The focus of this work is exclusively on the design of self-optimizing control structures where complexity increases with the number of disturbances. Thus, only the essential disturbance and operation scenarios are considered. The hope is that process non-linearities are small such that control strategies which turn out to be good for the nominal operating point operation do not completely fail for abnormal operation. Some disturbances which are not considered are fouling of heat exchangers, *e.g.*, caused by algae in water coolers or freeze out of HHCs in the precooler (Reithmeier *et al.*, 2004/04/21-24), feed variations in pressure/temperature/composition, *e.g.*, caused by upstream adsorber switching. A relatively common scenario which is not taken into account is turndown operation, for instance.

8. Besides reliability, the maximization of the operational benefit (profit function) is the most important objective for plant operation. The accurate solution of this problem is by far not a trivial task, as rigorous models of the plant (prediction of operational cost, operational constraints and production rate) and the market (sale price depends on supply and demand) would be required in order to make an appropriate decision. These decisions are usually taken on hierarchical layers as indicated in Figure 1.1. This work aims to find general control structures on the (low) regulatory control layer where "general" is meant in terms of independence on plant characteristics and market conditions. It is thus convenient to focus on maximization/minimization of simple profit/cost functions, *i.e.*, on maximizing the coefficient of performance (COP, Haywood, 1980, p. 75), maximizing the plant throughput subject to a given shaft power or minimizing the energy consumption subject to a given throughput. Note that these objectives are contradictory and may lead to totally different operating points. The findings in Appendix 4.A suggest that maximizing the LNG throughput is the most general objective independent of plant configuration and market conditions.

9. Superheating of the mixed refrigerant at compressor inlet must be provided in order

to prevent from droplet entry which may cause machinery damage and spoil safe operation (Singh and Hovd, 2006/09/28-29). The degree of superheating depends on temperature, pressure and composition and as all three are somewhat uncertain due to process noise, the degree of superheating cannot be reduced to zero in practice but must satisfy a defined safety margin T_{saf} (usually 10 K). However, from theory of thermodynamic cycles it can be concluded that minimum superheating is generally optimal for simple cycles in terms of maximum efficiency[1] (see limiting Carnot efficiency in the work of Haywood, 1980, p. 102). Jensen (2005/05/29-06/01) and Jensen and Skogestad (2007b,a) investigated the operation of simple heat pump and refrigeration cycles and came to the same conclusion. One way of keeping the degree of superheating at its lower limit is to fix all three influences. *E.g.*, composition is fixed by manual dosing, suction pressure is controlled by compressor regulation and suction temperature is controlled via throttle valve. In order to provide more flexibility in control structure design, another strategy than this is proposed. The dependency of the dew point temperature on the pressure can be approximated with a low order polynomial for the nominal composition and is represented by $T_{\mathrm{dew}}(p)$. The superheating is then kept at its safety margin by controlling the measure $(T_{\mathrm{dew}}(p) + T_{\mathrm{saf}})/T$. This approach require one MV less than the above stated and is referred to as minimum superheating control (SHC) throughout the thesis.

10. There are several operational limitations in LNG liquefaction processes which need to be maintained in order to ensure reliable operation of the plant. Override controllers in the regulatory layer are usually used to keep the plant away from these limitations, as plant shut down will be triggered owing to their violations. For instance, some protection mechanisms relating to compressor operation are shut down due to drop of suction pressure below its lower limitation or due to exceeding of shaft power limitation. In this work, considerations on operational limitations are made subsequently to the identification of self-optimizing control structures. For instance, compressor regulation which in the first instance remains unused for control purposes may later serve as an MV for keeping the compressor within normal operation mode.

From these points the following conclusions can be drawn for self-optimizing CSD for LNG liquefaction processes. The LNG temperature control using the LNG flow rate is an inherent control loop of the system and is not suspended. The compressor regulation is considered a feed-forward variable (not used as a controller MV). *I.e.*, compressor regu-

[1] It is important to stress that minimal subcooling must not be necessarily optimal for more complex cycles.

lation corresponds to a DV in the context of self-optimizing control. LNG throughput as the most general objective is to be maximized subject to a defined shaft power. As shaft power and compressor regulation are closely related, this problem it is almost equivalent to maximizing the LNG throughput subject to unaffected compressor regulation which is considered in the first place. The ambient temperature and the LNG setpoint temperature are considered the only DVs besides compressor regulation. No abnormal operation scenarios are taken into account.

4.4.2. Decision measures for feedback and advanced control

As stated by Engell *et al.* (2005/07/04-08), the effect of feedback control on the cost function in the presence of disturbances can be expressed as

$$\Delta J(\boldsymbol{d}) = J\left(\boldsymbol{u}_{\mathrm{S/O}}, \boldsymbol{d}\right) - J(\boldsymbol{0}, \boldsymbol{0}) = \\ \underbrace{J\left(\boldsymbol{u}_{\mathrm{S/O}}, \boldsymbol{d}\right) - J\left(\boldsymbol{u}_{\mathrm{R/O}}, \boldsymbol{d}\right)}_{=L(\boldsymbol{d})\geq 0} - \underbrace{\left(J(\boldsymbol{0}, \boldsymbol{d}) - J\left(\boldsymbol{u}_{\mathrm{R/O}}, \boldsymbol{d}\right)\right)}_{=L_0(\boldsymbol{d})\geq 0} + \underbrace{J(\boldsymbol{0}, \boldsymbol{d}) - J(\boldsymbol{0}, \boldsymbol{0})}_{=\Delta J_0(\boldsymbol{d})}.$$

For better comprehensibility, the terms L, L_0 and ΔJ_0 are visualized in the sketch in Figure 4.3. L is the convex loss (3.4) as defined by Halvorsen *et al.* (2003), *i.e.*, the difference between the optimal compensation of the disturbance (solid) and the compensation which is achieved by the chosen feedback control structure (dashed). The also convex second term L_0 represents the loss of leaving the MVs unaffected (dash-dotted). By using (3.3) and (3.6),

$$L_0 = \frac{1}{2}\boldsymbol{d}^T J_{\boldsymbol{ud}}^T J_{\boldsymbol{uu}}^{-1} J_{\boldsymbol{ud}}\, \boldsymbol{d}$$

can be derived. The non-convex third term ΔJ_0 is a measure of how the cost is affected by disturbances only. From (3.3),

$$\Delta J_0 = J_{\boldsymbol{d}}^T \boldsymbol{d} + \frac{1}{2}\boldsymbol{d}^T J_{\boldsymbol{dd}}\, \boldsymbol{d}$$

follows directly. In the following E_{wc} denotes the worst-case expectation operator. *I.e.*, $E_{\mathrm{wc}}(L) = L_{\mathrm{wc}}$,

$$E_{\mathrm{wc}}(L_0) = L_{\mathrm{wc},0} = \frac{1}{2}\bar{\sigma}^2\left(\left(J_{\boldsymbol{uu}}^{1/2}\right)^{-T} J_{\boldsymbol{ud}}\, \boldsymbol{W_d}\right) \tag{4.4a}$$

$$E_{\mathrm{wc}}(\Delta J_0) = \Delta_{\mathrm{wc}} J_0 = \left\|J_{\boldsymbol{d}}^T \boldsymbol{W_d}\right\|_1 + \frac{1}{2}\max\left|\lambda\left(\boldsymbol{W_d}\, J_{\boldsymbol{dd}}\, \boldsymbol{W_d}\right)\right| \tag{4.4b}$$

$$E_{\mathrm{wc}}(\Delta J) = \Delta_{\mathrm{wc}} J = \left\|J_{\boldsymbol{d}}^T \boldsymbol{W_d}\right\|_1 + \frac{1}{2}\max\left|\lambda\left(\boldsymbol{M}^T \boldsymbol{M} + \begin{bmatrix} \boldsymbol{X} & \\ & \boldsymbol{0}_{n_y \times n_y}\end{bmatrix}\right)\right| \tag{4.4c}$$

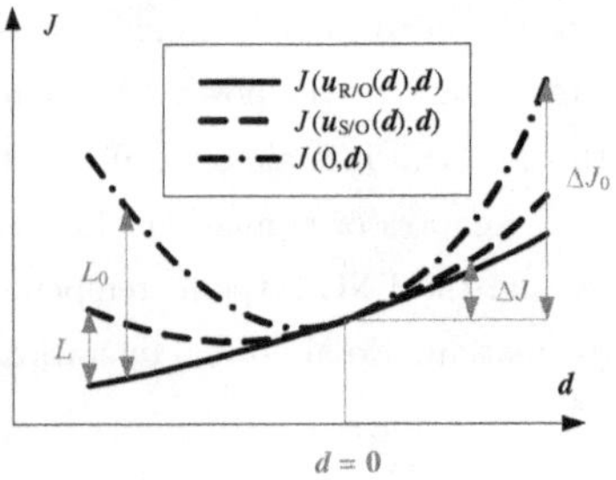

Figure 4.3.: Effect of feedback control on cost function and contributing terms

where $\boldsymbol{X} = \boldsymbol{W_d}\left(J_{\boldsymbol{dd}} - J_{\boldsymbol{ud}}^T J_{\boldsymbol{uu}}^{-1} J_{\boldsymbol{ud}}\right)\boldsymbol{W_d}$. The fact that the average expectation is not considered here is due to its more complex evaluation as it depends on the probability density function of the DVs and the implementation errors. Note that if both, the DVs and the implementation errors, are uniformly distributed, the average expectation can be evaluated as stated in the work of Kariwala *et al.* (2008, Proposition 1).

If $\Delta_{\mathrm{wc}}J_0 \gg L_{\mathrm{wc},0}$, or if L_{wc}, $L_{\mathrm{wc},0}$ and $\Delta_{\mathrm{wc}}J_0$ are all relatively small, then a variation of the manipulated variables offers no advantage, and neither optimization nor feedback control is required for this disturbances. If $\Delta_{\mathrm{wc}}J \gg L_{\mathrm{wc}}$ does not hold for a particular regulating control structure $\boldsymbol{H}$, then online optimization or an adaptation of the set-points should be performed rather than just regulation of the chosen variables to fixed precomputed setpoints.

4.5. Simple SMR cycle/C3MR process

In this section, the self-optimizing CSD for the simple SMR cycle and the propane pre-cooled mixed refrigerant (C3MR) cycle is presented. Figure 4.4 shows the process flow diagram (PFD) of the C3MR process with the T/FR/CRC control structure applied. Newton (1986/07/10), the inventor of this control structure, claimed that it achieves the highest production per unit of energy consumed. A controllability study by Mandler *et al.* (1997/07/24) revealed that it is less favorable in terms of dynamic considerations. The T/FR/CRC structure is compared to newly developed control structures based on average loss and feasible operating range.

For the sake of generalization, control structures are identified based on a model of the (simple) single mixed refrigerant (SMR) process. Since the SMR and the C3MR process differ only by the propane precooling cycle, the resulting control structures are applicable to both cycles. The process of the peakshaving LNG plant indicated in Figure 4.5 serve as a basis for this study. The plant was dynamically modeled in 1999 after Linde installed

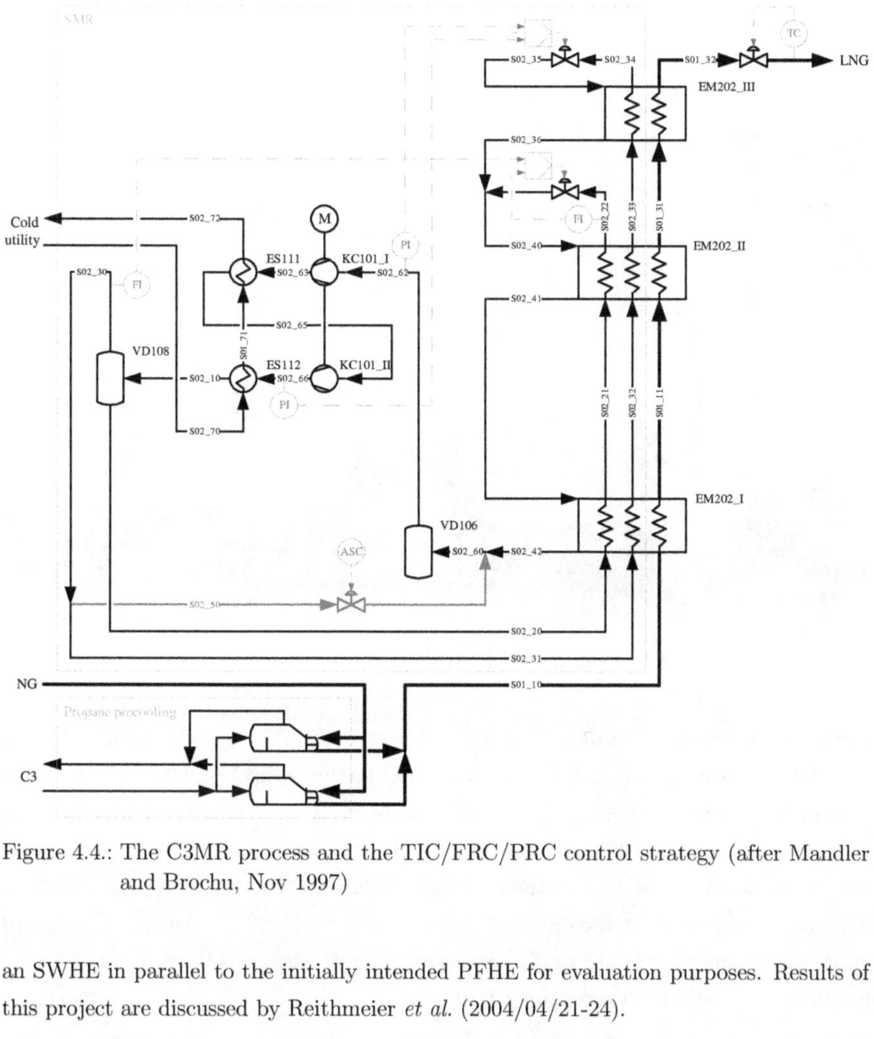

Figure 4.4.: The C3MR process and the TIC/FRC/PRC control strategy (after Mandler and Brochu, Nov 1997)

an SWHE in parallel to the initially intended PFHE for evaluation purposes. Results of this project are discussed by Reithmeier *et al.* (2004/04/21-24).

In the C3MR process, the LNG is precooled in a farm of kettle type heat exchangers which are fed with propane by a refrigeration cycle. The liquefaction and subcooling takes place in two or three subsequent SWHEs where light and heavy mixed refrigerant (LMR and HMR, respectively) are used as the coolant. Figure 4.5 shows three SWHE in series because in the Mossel Bay plant HHCs needed to be separated from the methane-rich stream after the EM202_I. This separation was disregarded in order to achieve a maximum degree of freedom for optimization. The SMR cycle consists of a two-stage compressor (KC101_I and KC101_II) with inter- and aftercooler (ES111 and ES112). In the ES112, the mixed refrigerant is partly condensed. The light and the heavy fraction are separated in the separator VD108. The level in the VD108 is not controlled as it

Figure 4.5.: Peakshaving LNG plant Mossel Bay, South Africa

carries the excessive refrigerant. The LMR after the VD108 is liquefied on the tube side of EM202_I and EM202_II and then subcooled on the tube side of EM202_III. Afterwards it is throttled and serves as the shell side coolant of all SWHEs. The HMR after the VD108 is subcooled on the tube side of the EM202_I and EM202_II and then throttled and mixed to the LMR to give the shell side coolant for the EM202_I and EM202_II. The mixed refrigerant at the outlet of the shell of the EM202_I is usually superheated (with a considerable distance to the dew point). Nevertheless, safety reasons require the installation of the droplet separator VD106.

4.5.1. Degree of freedom analysis

The SMR cycle in the configuration of Figure 4.4 has $n_c + 2$ degrees of freedom defined in the sequel.

1 Compressor speed n_r. Note that in a single shaft configuration, one speed applies to both stages.

2 Positions of the warm and cold Joule–Thomson (WJT/CJT) valves. They indirectly affect pressure levels, heavy and light mixed refrigerant flow rates and the active

charge[2]. Instead of the valve positions, the flow rates of the HMR and LMR, F_{HMR} and F_{LMR}, are considered as MVs. *I.e.*, it is supposed that flow controllers manipulate the valve positions.

$n_c - 1$ Mixed refrigerant composition

Note that in some plants the compressor inter- and aftercooler have adjustable cooling capacity in order to keep the mixed refrigerant outlet at constant temperature irrespective of the ambient (water/air) temperature. This is desirable as it stabilizes the cycle and decouples it from short-term ambient temperature variations. This introduces extra degrees of freedom which exclusively serve as MVs of the mixed refrigerant temperature controllers. For a more detailed degree of freedom analysis of the C3MR process, the reader is referred to the work of Jensen and Skogestad (2009c).

4.5.2. Model setup

The CSD procedure for the SMR cycle was performed according to the workflow introduced in Section 4.2. A dynamic simulation model of Mossel Bay plant in OPTISIM® was already available from a former operation study and equals the SMR part of the flowsheet in Figure 4.4. For the sake of doing steady state studies, the model was slightly changed such that the absolute charge of each species remained fixed in the cycle. *I.e.*, the model was rebuilt in quasi-closed configuration as pointed out in Section 2.5.1. It is important to stress that only the inventories in the drums were taken into account as they carry the major charge.

Two objectives, the COP and the LNG throughput F_{LNG}, were optimized by variation of the MVs, F_{HMR} and F_{LMR}, subject to an ambient temperature of $T_{\text{amb}} = 299.15\,\text{K}$, the compressor rating speed ($n_{\text{r}} = 10927\,\text{RPM}$), an LNG setpoint temperature of $T_{\text{LNG}}^{\text{sp}} = 115\,\text{K}$, as well as a lower limit on the suction pressure ($p_{\text{suc}} \geq 3.5\,\text{bar}$), the superheating ($\Delta T_{\text{SH}} \geq 10\,\text{K}$) and the compressor flow (surge line specified by vendor). The species' inventories within the cycle might be used as optimization variables but were kept at design conditions. The resulting optimal operating points for COP and LNG throughput maximization do not coincide and are given in Table 4.4. None of the imposed constraints are active at the optimal operating points. Note that the objective values for both optima are also indicated in the chart in Figure 4.15 for illustration purposes. Due to the argumentation given in Appendix 4.A, the throughput was selected as the objective for self-optimizing control and the nominal operating point was selected at maximum throughput conditions. The operating ranges of the input variables at this nominal point were

[2]According to Jensen and Skogestad (2007b), the active charge refers to the total mass in the cycle except for the mass in the buffer tank with variable level. One (steady-state) degree of freedom is lost if the level is fixed and dosing/venting is prevented.

	max COP	max F_{LNG}
COP (in %)	54.72	50.31
F_{LNG} (in mol/s)	203.2	241.1
F_{HMR} (in mol/s)	181.8	216.1
F_{LMR} (in mol/s)	368.8	539.9
p_{suc} (in bar)	3.6	4.4
ΔT_{SH} (in K)	41.6	40.3
ΔF_{surge} (in mol/s)	29.3	128.0

Table 4.4.: Nominal values at optimal operating points of the Mossel Bay plant

	Lower bound	Nominal point	Upper bound
F_{HMR} (in mol/s)	195 (−16.7 %)	216.0	255 (+15.7 %)
F_{LMR} (in mol/s)	460 (−14.8 %)	539.97	620 (+14.8 %)
n_{r} (in RPM)	10700 (−2.1 %)	10927	11400 (+4.3 %)
$T_{\text{LNG}}^{\text{sp}}$ (in K)	113.5 (−1.3 %)	115	117.5 (+2.2 %)
T_{amb} (in K)	284.15 (−5.0 %)	299.15	305.15 (+2.0 %)

Table 4.5.: Operating range of the Mossel Bay plant

evaluated. *I.e.*, a variation of each MV/DV was performed in upper and lower direction until either a reasonable distance to the nominal point was reached or until convergence failed indicating the edge of the feasibility region. The results are shown in Table 4.5. As suction pressure control by compressor regulation was disregarded, the compressor speed was considered a DV.

4.5.3. CSD

The MV and DV vectors are respectively given by $\boldsymbol{u} = \begin{bmatrix} F_{\text{HMR}} & F_{\text{LMR}} \end{bmatrix}^T$, *i.e.*, the HMR and LMR flow rate, and $\boldsymbol{d} = \begin{bmatrix} T_{\text{amb}} & T_{\text{LNG}}^{\text{sp}} & n \end{bmatrix}^T$, *i.e.*, the ambient temperature, the LNG setpoint temperature and the compressor speed. The PV vector $\boldsymbol{y}$ consists of 18 variables indicated in Table 4.6. The profit function is the LNG throughput, *i.e.*, $J = F_{\text{LNG}}$. According to these specifications, the steady-state I/O model and the Hessian at the nominal point can be obtained. It is worth noting that the Hessian was calculated by permutation of MVs and DVs over the ranges given in Table 4.5 and using a least squares fit of (3.3) on the resulting LNG throughput. The scaling matrix for the disturbances follow the permutation ranges in Table 4.5, *i.e.*, $\boldsymbol{W_d} = \text{diag}\,(10\,\text{K}, 1.5\,\text{K}, 350\,\text{RPM})$. For the scaling matrix representing the implementation error, $\boldsymbol{W_{n^y}}$, it was assumed that 1 % flow uncertainty, 0.5 K absolute temperature uncertainty and 10 mbar absolute pressure uncertainty are present.

With the presented information, commonly known control structures were judged in terms of expected worst-case/average loss. The results are presented in the first three

Stream	F	T	p
S01_11	1	5	
S01_31		6	
S02_10	2	7	17
S02_21	3	8	
S02_31	4	9	
S02_22		10	
S02_32		11	
S02_34		12	
S02_36		13	
S02_41		14	
S02_42		15	18
S02_70		16	

Table 4.6.: Measurement and their indices in $\boldsymbol{y}$ for the SMR cycle

lines of Table 4.7. Also indicated is $\Delta_{\text{wc}}J$, the worst-case effect of feedback on cost (4.4c). For better illustration, all values are related to the nominal throughput and are given in %. The case indicated by MV0 refers to open loop configuration and unaffected MVs. As the implementation error is negligible, the worst-case loss of MV0 equals the measure $L_{\text{wc},0}$ defined in (4.4a) and $\Delta_{\text{wc}}J$ of MV0 corresponds to $\Delta_{\text{wc}}J_0$ defined in (4.4b). As the $L_{\text{wc},0}$ and $\Delta_{\text{wc}}J_0$ are not small and $L_{\text{wc},0} \ll \Delta_{\text{wc}}J_0$ does not hold, feedback control can be reasonably applied for loss reduction. The conventional control structure of the Mossel Bay plant indicated by CONV shows moderate loss. Note that the suction pressure control which is actually included in the CONV structure was disregarded in order to make the comparison of the worst-case/average loss possible. The T/FR/CRC structure as shown in Figure 4.4 is indicated in Table 4.7 in its linearized formulation. It turned out that it is a very uneconomic control structure in terms of worst-case/average loss although it was claimed that it produces the highest LNG production per unit of energy consumed (Newton, 1986/07/10). Note that losses above 100 % as in the T/FR/CRC case are meaningless but suggest that the control structure has a small feasibility region.

The best control structure in terms of worst-case loss subject to PV selection was found by application of the B3WC method and was named SEL. Its performance in terms of worst-case/average loss is better than MV0 by a factor of 1/3. COM2 indicates the best common-sized control structure with PV subset size two and pure unit combinations found via application of the MIAV method. Temperature combinations were disregarded as they cannot be physically interpreted. Due to the small system dimensions, the application of the MIAV method was not expensive. In comparison to SEL the worst-case/average loss of COM2 is further decreased by one order of magnitude. IND was obtained similarly to COM2 with the difference that no flow rate combinations were allowed. Its worst-

Name	Linear control structure	L_{av} (in %)	L_{wc} (in %)	$\Delta_{wc}J$ (in %)
MV0	$\{F_{HMR}, F_{LMR}\}$	0.43	6.38	38.25
CONV	$\{T_{S01\_31}, F_{LMR}\}$	1.09	16.37	48.23
T/FR/CRC	$\left\{\begin{array}{l} F_{LMR} - 1.02\, F_{HMR}, \\ p_{S02\_10} - 9.89\, p_{S02\_42} \end{array}\right\}$	178.5	4284.5	4315.2
SEL	$\{F_{LMR}, T_{S02\_42}\}$	0.15	2.17	33.95
COM2	$\left\{\begin{array}{l} F_{LNG} - 1.19\, F_{MR}, \\ F_{LMR} - 1.41\, F_{HMR} \end{array}\right\}$	0.01	0.19	31.97
IND	$\left\{\begin{array}{l} T_{S02\_42}, \\ p_{S02\_10} + 9.36\, p_{S02\_42} \end{array}\right\}$	0.05	0.95	32.07
HHC/FRC	$\left\{\begin{array}{l} T_{S01\_31}, \\ F_{NG} - 0.31\, F_{HMR} - 0.73\, F_{LMR} \end{array}\right\}$	0.85	20.18	51.69

Table 4.7.: Worst-case/average loss of linear control structures for the SMR cycle

case/average loss is thus only half as good as COM2.

For the SEL, COM2 and IND structure, it holds the relationship $L_{wc} \ll \Delta_{wc}J$. Accordingly, if the measures L_{wc} and $\Delta_{wc}J$ in Table 4.7 are really representative for the nonlinear behavior, it can be concluded that feedback control by these structures is almost optimal and optimization of controller setpoints, for instance by the use of RTO, can be considered redundant.

4.5.4. Nonlinear verification

The nonlinear steady-state behavior of control structures indicated in Table 4.7 have been investigated by parameter studies of the DV set within the ranges given in Table 4.5. The main results are presented in Figure 4.6. The charts in the first row represent the LNG throughput vs. the three DVs. The R/O case represents the best achievable behavior and was obtained via the optimization functionality of OPTISIM®. The R/O curve thus indicates the upper bound of all other curves which are related to feedback control structures. As the resolution is fairly poor, the deviation between the R/O curve and all other curves is respectively plotted in the diagrams in the second row. No operability constraints were violated over the observed DV range. Expectedly, the maximum achievable LNG throughput (R/O) is monotonically decreasing with rising ambient temperature, falling LNG setpoint temperature and falling compressor speed. Curve ends indicate the edge of the feasible operating range.

Considering ambient temperature drop, the worst-case loss of the MV0 structure underestimates the actual behavior. The curves of the CONV structure meet its worst-case loss value relatively well. It is more favorable than the MV0 policy. As predicted by its worst-case/average loss, the T/FR/CRC structure indicated in Figure 4.4 shows very

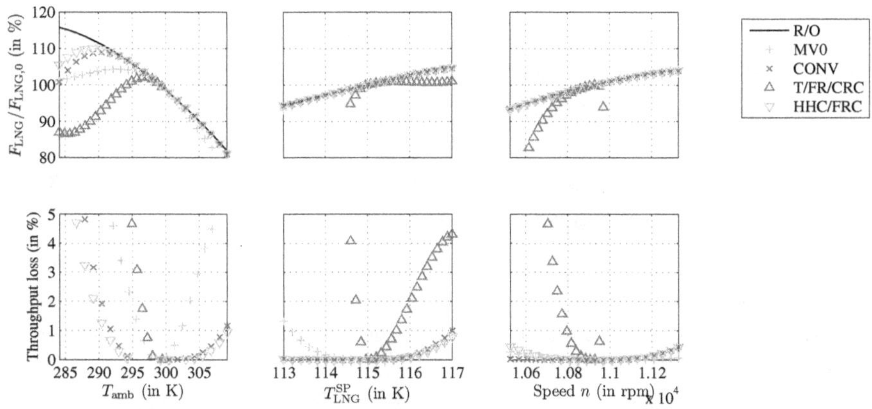

Figure 4.6.: Results of nonlinear verification of control structures for the SMR process

undesirable performance.

Remark 4.2. It is striking that the feasibility of the T/FR/CRC structure is violated on either side of the domain of each of the three DVs. This effect is due to unfavorable combination of CVs which is pointed out here. It is important to notice that by holding the F_{LMR}/F_{HMR} ratio constant, the vapor fraction of the partly condensed mixed refrigerant is fixed and, thus, its pressure, *i.e.*, the discharge pressure, depends on the ambient temperature. Accordingly, if the ambient temperature rises, the discharge and suction pressure increase as their ratio must remain invariant. However, the suction pressure level is bounded from above as it influences the evaporation temperature of the refrigerant which is already close to the condensation temperature of the tube-side fluids at nominal conditions. The same argumentation reversed can be applied in order to give an explanation for the feasibility violation of the LNG setpoint temperature. A decrease of the LNG setpoint temperature requires a drop of the temperature level of the evaporating refrigerant which can only be achieved by suction pressure reduction. This, however, is not possible due to invariant pressure ratio and fixed ambient temperature and discharge pressure. The reason that the compressor speed is bounded from above is more profound and a proper explanation cannot be given.

Remark 4.3. The application of the T/FR/CRC structure to the C3MR process may be less severe as suggested by the curves in Figure 4.6, as the ambient temperature disturbance can be rejected by the propane precooling cycle. However, the small feasibility region in terms of LNG setpoint temperature is likely to occur anyway. These results verify comments from customers which state that the T/FR/CRC policy is difficult to implement (Mandler *et al.*, 1997/07/24).

The nonlinear performance of the control structures SEL, COM and IND did not meet the expectations raised by their worst-case/average loss figures. All of them failed against the CONV structure in terms of feasibility range or worst-case loss or both. Via a trial and error approach, a promising self-optimizing control structure was found, given by $\{T_{\mathrm{S01\_31}}, F_{\mathrm{NG}}/(F_{\mathrm{HMR}} + F_{\mathrm{LMR}})\}$. It is indicated as HHC/FRC in Table 4.7 (in its linearized formulation). The HHC/FRC structure performs only slightly better than CONV but has the downside that it is not practically proven.

It can be generally concluded that the values of worst-case/average losses are not always well in agreement with the nonlinear behavior. The conventional control structure implemented in Mossel Bay has good self-optimizing abilities and could not be outperformed by any other control structure.

4.6. LIMUM® cycle

This section deals with the self-optimizing CSD for the (proprietary) Linde-Multi-Stage-Mixed-Refrigerant (LIMUM®, Stockmann *et al.*, 1997/05/28) cycle. It is a proven technology (Berger *et al.*, 2003/06/01-05) for the liquefaction of natural gas in small to medium-sized plants, *i.e.*, plants with less than 3 MTPA. It is a single mixed refrigerant cycle and its process topology is a deviation from that of the simple SMR cycle treated in Section 4.5. The LIMUM® cycle will be incorporated in one of the first LNG floating production storage and offloading units (FPSO) developed by SBM Offshore and Linde AG. The first vessel indicated in figure 4.7 is planned to be delivered by 2012 (Voskresenskaya, 2009/03/31-04/02). It is designed for gas fields with reserves of 1-4 TCF and liquefaction capacity of 2.5 MTPA LNG plus condensate and LPG production.

Figure 4.8 shows a PFD of the LIMUM® cycle applied in an LNG liquefaction plant in Stavanger, Norway going on stream in 2010. The cooling and condensation of the natural gas stream S1_101 takes place in three serial SWHE bundles, E11, E12 and E13 which correspond to precooling, liquefaction and subcooling, respectively. A mixed refrigerant of the typical components nitrogen, C1, C2 and C4 is used. At the first pressure stage, the mixed refrigerant stream S1_202 is compressed from the low to the medium pressure by the compressor C02A and partially condensed by the aircooler E21. The partly condensed stream S1_204 is separated into a liquid and a vapor stream by the separator D23. The liquid stream S1_221 represents the medium pressure heavy mixed refrigerant (MHMR) and serves as the coolant for the precooling bundle E11. At the second pressure stage, the medium pressure vapor stream S1_215 is compressed and cooled by the compressor C02B and the aircooler E23, respectively. The high pressure vapor stream S1_232 is partly condensed and separated in the warm bundle E11 and the separator D12, respectively.

Figure 4.7.: LNG FPSO unit

This gives the high pressure heavy mixed refrigerant (HMR) stream S1_141 which serves as the coolant for the liquefaction bundle E12 and the high pressure light mixed refrigerant stream (LMR) which is taken as the subcooling coolant. It is important to stress that no HHC removable is implemented after the precooler.

The regulatory control strategy implemented in the Stavanger plant is also indicated in the PFD in Figure 4.8. The suction pressure is controlled by manipulation of the inlet guide vane (IGV) position of the first stage. The LMR flow rate is fixed by manipulating the CJT valve. The natural gas temperature after the precooler is controlled by manipulating the WJT valve. However, this loop is overridden if the temperature of the mixed refrigerant before the compressor drops below a lower limit.

4.6.1. Degree of freedom analysis

In general, the topology deviation between the LIMUM® and the simple SMR cycle has no impact on the operational degrees of freedom. This is due to the fact that there are as many separators as throttle valves introduced and their levels need to be stabilized by control loops. As in the SMR, the level of one separator remains uncontrolled as it is an inherently stable mode and allows for a variable active charge in the cycle (Jensen and

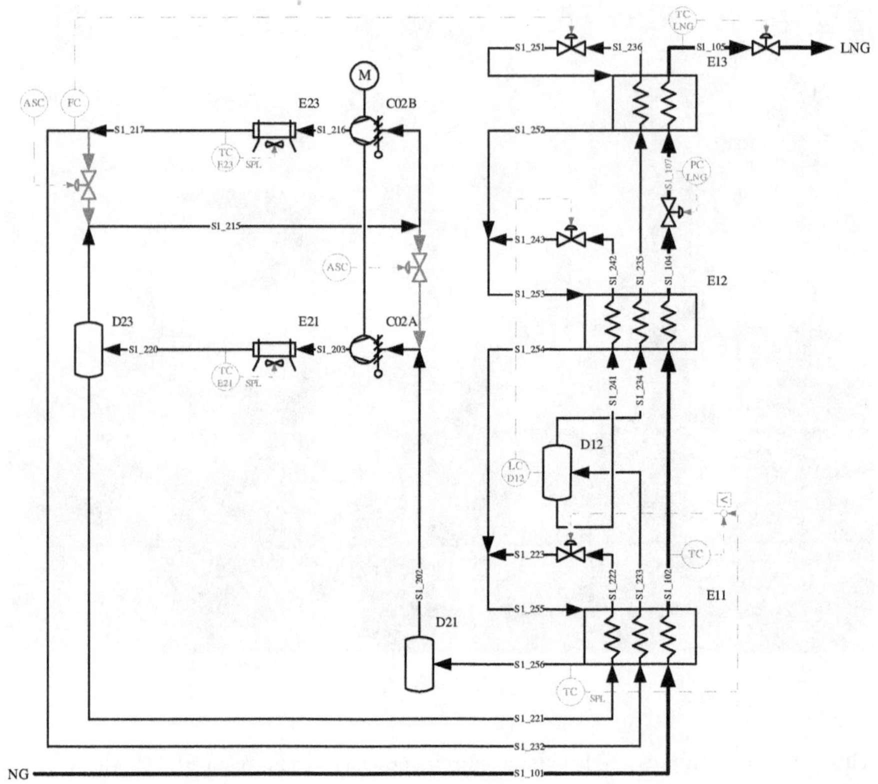

Figure 4.8.: LIMUM® cycle with regulatory control structure of the Stavanger plant

Skogestad, 2007b).

Some additional features of the process in Figure 4.8 are that (i) throttling of the LNG takes place before the subcooler E13, (ii) IGV compressor regulation is used and (iii) aircoolers are applied. All three introduce further degrees of freedom. However, the impact of the first on process operation can be neglected. Regarding the second it is important to notice that two-stage single-shaft compressors can include two IGV rows, however only in back-to-back configuration (Lüdtke, 2004, p. 119). The third point relates to the fact that aircoolers are usually equipped with vans and their speed can be used to regulate the outlet temperature of the partly condensed mixed refrigerant which can be convenient for reducing the interference of the ambient temperature on the liquefaction process. This means that during times when ambient temperature is low, the fans' speeds are reduced by the controllers TCE21 and TCE23 and vice versa. Note that the aim of small interference does not generally conflict with the objective of high efficiency as shown below.

Accordingly, the LIMUM® cycle in the configuration of Figure 4.8 has $n_c + 5$ degrees of freedom available as MVs:

2 IGV positions z_1, z_2

1 MHMR flow rate F_{MHMR} (or position of relating throttle valve)

1 LMR flow rate F_{LMR} (or position of relating throttle valve)

2 Fan speeds of aircoolers (or setpoints of temperature controllers)

$n_c - 1$ Mixed refrigerant composition

4.6.2. Model setup

The basis of the CSD was an OPTISIM® design flowsheet of the Stavanger plant. The transformation from the design into a dynamic simulation flowsheet was carried out as described in Section 4.2. For the sake of steady-state investigations, the cycle needed to be modeled in quasi-closed configuration in order to force the species' inventories within the cycle to remain invariant (as discussed in Section 2.5.1). Thereby, only the inventories of the drums were taken into account as they include the major refrigerant mass (approx. 80 %).

The COP and the LNG throughput were optimized by variation of the MVs, F_{MHMR} and F_{LMR}, subject to an ambient temperature of $T_{amb} = 296\,K$, an LNG setpoint temperature of $T^{sp}_{LNG} = 111.66\,K$, IGV angles of $z_1 = z_2 = 20\,°$, as well as operating constraints for suction pressure ($p_{suc} \geq 3.5\,bar$), superheating at compressor inlet ($\Delta T_{SH} \geq 10\,K$) and compressor surge. The results for maximum LNG throughput and maximum COP are shown in Table 4.8. The objective values are additionally indicated in Figure 4.15 for illustration purposes. The superheating constraint is active for both optima. The operating point for the maximum COP case is located closer to the surge line than the maximum LNG throughput point. This effect was already observed for the simple SMR cycle and is reasonable as the compressor operation is usually very efficient close to the surge line as exemplarily indicated by the characteristic compressor maps in Figure 4.9a and b. The surge margin for the maximum COP case is very small and suggests that the process is optimally active at the surge line somewhere within the operating region. An analogical result was obtained by Jensen and Skogestad (2009a) for the PRICO process.

Remark 4.4. The ambient temperature at nominal point was selected offset from design conditions at $T_{amb} = 296\,K$ where a climax of the maximum LNG throughput was observed as indicated in Figure 4.10. Due to this effect, it is economically reasonable to

(a) Speed regulation (b) Adjustable IGV regulation

Figure 4.9.: Characteristic maps of a centrifugal compressor (from Lüdtke, 2004, pp. 116,119)

	max COP	max F_{LNG}
COP (in %)	81.09	72.18
F_{LNG} (in mol/s)	535.2	644.1
F_{MHMR} (in mol/s)	339.1	410.9
F_{LMR} (in mol/s)	451.8	621.7
p_{suc} (in bar)	4.1	4.0
ΔT_{SH} (in K)	10.0	10.0
$\Delta F_{\mathrm{surge}}$ (in mol/s)	45.19	423.0

Table 4.8.: Nominal values at optimal operating points of the Stavanger plant

control the temperature of the partly condensed mixed refrigerant via speed manipulation of fans attached to the aircoolers. If $T_{\mathrm{amb}} \geq 296\,\mathrm{K}$, the fans should be run at full speed, otherwise, temperature control should be active.

Remark 4.5. The effect that the maximum LNG throughput versus ambient temperature shows a climax as indicated in Figure 4.10 is an unexpected and puzzling outcome. It is rather expected that the refrigeration capacity declines monotonically with ambient temperature as can be observed for the most refrigeration cycles such as the ammonia cycle investigated by Jensen and Skogestad (2007a) or the simple SMR cycle treated in Section 4.5. The best explanation for the counterintuitive behavior is based on the fact that the ambient temperature affects, on the one hand, the temperature level of the cold and warm T-Q composite curves and, on the other hand, their slopes as a result of the variation of the mixed refrigerant composition. On the left hand side of the climax, it seems that the slope of the cold curve changes so unfavorable that it overcompensates its drop in temperature level. An interesting question is why this effect cannot be observed for the simple SMR cycle. In the simple SMR cycle there are two refrigerants, the HMR and the LMR, whose flow rates can be manipulated independently from each other providing the ability to adapt the cold to the warm T-Q composite curve via two degrees of freedom. In the LIMUM® cycle there are on the one hand three refrigerants, MHMR for precooling, HMR for liquefaction and LMR for subcooling,

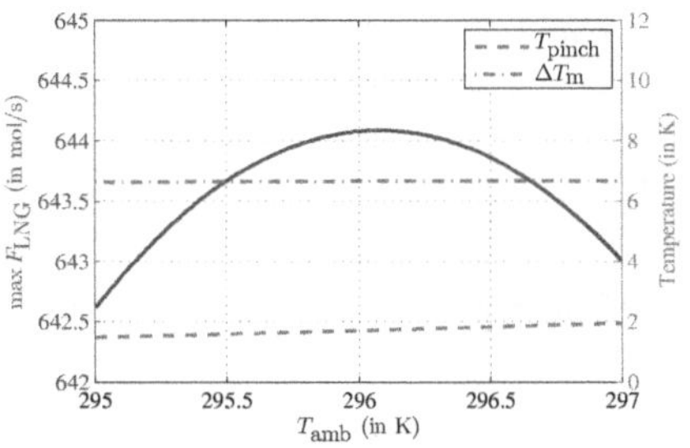

Figure 4.10.: Climax of maximum LNG throughput

separating the cold T-Q composite curve into three segments. On the other hand, the three flow rates cannot be manipulated independently from each other as only one degree of freedom, *i.e.*, the LMR flow, is present (IGV angles are fixed, minimum superheating is satisfied). One degree of freedom for adapting the slope of the three segments in the cold T-Q composite curve is too few for compensation of their individual variation due to composition change. This conjecture can be proven by the observation of the pinch temperature T_{pinch}, *i.e.*, the smallest temperature difference in all SWHEs, and the mean temperature difference over all SWHEs ΔT_m in Figure 4.10. A small pinch temperature is forbidden due to increasing need of exchange area per heat transfer unit. As the heat exchange area is fixed, a smaller pinch temperature can only be obtained if elsewhere the temperature difference is increased. Due to the monotonically falling pinch temperature in direction of smaller ambient temperature together with a slightly decreasing mean temperature difference, at some point a further increase of the LNG throughput is inhibited.

As stated in Appendix 4.A, the maximum LNG throughput point can be considered generally optimal and was thus selected as the nominal operating point. The MHMR flow rate is used as an MV for satisfaction of minimum superheating of 10 K. Therefore, only one MV is left for self-optimizing CSD. Suction pressure control was disregarded and, thus, the IGV positions were considered DVs. The feasibly region was detected by parameter studies of input variables. The ranges are indicated in Figure 4.9 and were found to be fairly small, especially for the LNG setpoint temperature and the ambient temperature. The IGV angles were not investigated independently from each other but were synchronized, *i.e.*, $z_{1,2} = z_1 = z_2$. At either bound of each range an optimization of the LNG throughput was performed in order verify that the minimum superheating is everywhere optimally active which turned out to be indeed the case.

	Lower bound	Nominal point	Upper bound
F_{LMR} (in mol/s)	590 ($-5.1\,\%$)	621.7	650 ($+4.6\,\%$)
$z_{1,2}$ (in °)	-20 ($-200\,\%$)	20	40 ($+100\,\%$)
$T_{\mathrm{LNG}}^{\mathrm{sp}}$ (in K)	111.24 ($-0.38\,\%$)	111.66	112.66 ($+0.9\,\%$)
T_{amb} (in K)	294 ($-0.68\,\%$)	296	298 ($+0.68\,\%$)

Table 4.9.: Perturbation range of the Stavanger plant

Stream	F	T	p
S1_102	1	6	
S1_104		7	
S1_204	2	8	18
S1_217		9	19
S1_222	3	10	
S1_233		11	
S1_235	4	12	
S1_242	5	13	
S1_236		14	
S1_252		15	
S1_254		16	
S1_256		17	20

Table 4.10.: Measurement and their indices in $\boldsymbol{y}$ for the LIMUM® cycle

4.6.3. CSD

The MV vector is only represented by the LMR flow rate, *i.e.*, $\boldsymbol{u} = F_{\mathrm{LMR}}$. The DV vector is given by $\boldsymbol{d} = \begin{bmatrix} z_{1,2} & T_{\mathrm{LNG}}^{\mathrm{SP}} & T_{\mathrm{amb}} \end{bmatrix}^T$, *i.e.*, the (synchronized) IGV angles, the LNG set-point temperature and the ambient temperature, respectively. The PV vector $\boldsymbol{y}$ consists of 20 variables indicated in Table 4.10. The profit function is the LNG throughput, *i.e.*, $J = F_{\mathrm{LNG}}$. According to these specifications, the steady-state I/O model and the Hessian at the nominal point were obtained whereas the latter was calculated by permutation of MVs and DVs over the ranges given in Table 4.9 and using least squares fit of (3.3) on the resulting LNG throughput. The scaling matrix for the disturbances are specified as $\boldsymbol{W_d} = \mathrm{diag}\,(30\,°, 1\,\mathrm{K}, 10\,\mathrm{K})$. For the scaling matrix representing the implementation error, $\boldsymbol{W_{n^y}}$, it was assumed that 1 % flow uncertainty, 0.5 K absolute temperature uncertainty and 10 mbar absolute pressure uncertainty are present.

With the presented information, two a priori known control structures, MV0 and CONV, were judged in terms of expected worst-case/average loss. MV0 refers to unaffected F_{LMR}. CONV relates to the control structure indicated in Figure 4.8. The results of the worst-case/average loss and the measure $\Delta_{\mathrm{wc}}J$ are presented in Table 4.11. For better illustration, the values are related to the nominal throughput and are given in %. The loss of the MV0 structure is higher than in the simple SMR cycle but still mod-

Name	Linear control structure	L_{av} (in %)	L_{wc} (in %)	$\Delta_{wc}J$ (in %)
MV0	$\{\Delta T_{SH}, F_{LMR}\}$	7.23	86.81	142.45
CONV	$\{T_{S1\_102}, F_{LMR} + F_{HMR}\}$	24.18	360.62	391.61
SEL	$\{\Delta T_{SH}, F_{MR}\}$	0.46	5.58	59.83
COMB	$\{\Delta T_{SH}, F_{LNG} - 1.15\, F_{MR}\}$	2.6e−3	3.9e−2	59.81
HHC	$\{\Delta T_{SH}, T_{S1\_102}\}$	3.93	47.12	96.32

Table 4.11.: Worst-case/average loss of linear control structures for the LIMUM® cycle

erate. The worst-case loss of MV0 equals the measure $L_{wc,0}$ defined in (4.4a). $\Delta_{wc}J$ of the MV0 structure corresponds to $\Delta_{wc}J_0$ defined in (4.4b). As $L_{wc,0}$ and $\Delta_{wc}J_0$ are both fairly large and as $L_{wc,0} \ll \Delta_{wc}J_0$ is false, feedback control can be reasonably applied for loss reduction. The CONV structure has a considerable large loss and the fact that the worst-case loss is way above 100 % indicates a high likelihood of infeasibility over the perturbation range. *I.e.*, for some disturbance perturbations, it may happen that setpoint deviation occurs.

The SEL structure in Table 4.11 relates to the PV selection structure with least worst-case loss and was obtained by applying the B3WC method. It shows significant smaller loss than MV0 whereas the only difference is that F_{MR} is held fixed instead of F_{LMR}. Note that F_{MR} refers to the flow rate of S1_102. The worst-case/average loss can be further reduced to a considerable small worst-case/average loss by taking PV combinations into account. The COMB structure has the least worst-case loss of all control structures subject to combination of two PVs. It was found by application of the PB3WC method.

For all the so far considered structures in Table 4.11, only the COMB structure satisfies the relationship $L_{wc} \ll \Delta_{wc}J$, which indicates that optimization by feedback control achieves good performance and online optimization of setpoints, for instance by RTO, is expendable.

4.6.4. Nonlinear verification

The worst-case and average losses of all the control structure listed in Table 4.11 were judged by implementing them into the steady-state model of the Stavanger plant and investigating their nonlinear behavior. Switching of model equations occurred at nominal point due to setpoint low control of the temperatures after the aircoolers and override control present in the CONV structure. The results are presented in Figure 4.11. The first row of charts gives the LNG throughput versus the DVs ambient temperature, LNG setpoint temperature and (synchronized) IGV angles. The best possible behavior (uppermost curve) is indicated by R/O and was obtained by application of the optimization functionality of OPTISIM®. Below 296 K, LNG throughput is invariant with respect to

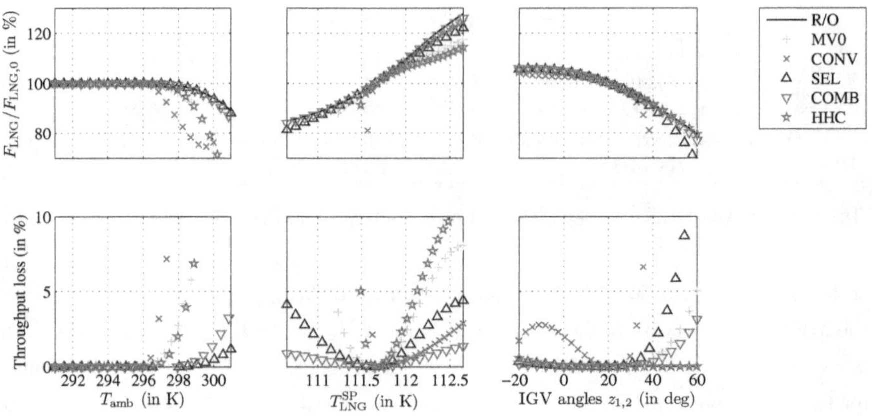

Figure 4.11.: Results of nonlinear analysis of control structures for the LIMUM® cycle

ambient temperature as the temperature of the mixed refrigerant is kept at fixed value via fan speed regulation. Above, 296 K, the maximum LNG throughput drops with rise of ambient temperature. Maximum LNG throughput is favored by high LNG setpoint temperatures and small IGV angles. The relative difference between the R/O curve and any of the lower curves denotes the loss. It is indicated in the second row of charts. Ending curves indicate that the edge of the feasibility region is reached.

The worst-case loss results of the MV0, the CONV and the SEL structure can be verified relatively well. It is important to stress that in case of reduced feasibility region, the worst-case loss can only be estimated via imaginary extrapolation of the respective curves. The COMB structure is admittedly small but not as tiny as suggested by the linear results.

Keeping the temperature of the natural gas fixed after the precooler is sometimes mandatory due to separation of HHCs in a vapor-liquid separator. This is only partly achieved by the CONV structure as the natural gas temperature controller is overridden by the SHC. It is thus suggested replacing the CONV structure by the HHC structure indicated in Table 4.11 if removal of HHCs after the precooler is necessary. As a consequence, a higher reliability for the achievement of product specification can be obtained. As can be seen by the loss figures and the nonlinear behavior in Figure 4.11, the economic performance is similar for both, the CONV and the HHC structure.

Figure 4.12.: Hammerfest LNG plant under construction at Melkøya Island, Barents Sea on April 2005

4.7. MFC® process

The Mixed Fluid Cascade (MFC®, Förg *et al.*, 2000/03/24) process was developed within the scope of the LNG technology alliance between Linde and Statoil. The first technical realization has taken place in a large scale baseload LNG liquefaction plant at Melkøya Island with gas from the Snøhvit field (Figure 4.12). The plant is named after the nearby city Hammerfest, Norway. Hammerfest is the northernmost LNG plant with very rough climate conditions. One of the reasons why the MFC® process was applied is that it is especially designed for cold climate. Beside the climate, there are several other challenges which have been tackled within the scope of this project (Heiersted, 2002/10/13-16; Berger *et al.*, 2003a; Bauer, 2005).

The development of a self-optimizing control structure for the MFC® process was based on a design study for the evaluation of the largest train size possible for warm climate called Cigma. The related PFD is shown in Figure 4.13. The natural gas is precooled, liquefied and subcooled in the three sequential SWHEs, 23E_02, 23E_03 and 23E_04, respectively. The refrigeration capacity of the shell-side stream of each SWHE is provided by a separate refrigeration cycle, the precooling cycle (PC), the liquefaction cycle (LC) and the subcooling cycle (SC) which are installed in cascade configuration. All three cycles use mixed refrigerant consisting of components from the set nitrogen and C1 through C4. The PC, LC and SC mixed refrigerant carry components from the right, from the middle and from the left of this spectrum, respectively. The compressor of each cycle is driven by a GE Frame 9 gas turbine. The SC has a two-stage intercooled compressor. For the LC, the intercooling is omitted due to construction issues. The PC has only one compressor stage. Aircoolers are used for transferring the heat to the ambient. To protect the compressors from surge, anti-surge bypasses are provided around each compressor. Figure 4.13 shows also a regulatory control structure equivalent to that implemented in the Hammerfest plant. The control strategy is similar among the cycles. In each cycle, the suction pressure is controlled by compressor speed regulation and the pressure before the throttling/expansion is fixed via JT valves. Another analogy is that in each cycle it is prevented that superheating of the mixed refrigerant at compressor inlet drops below a lower limit. It is in each cycle achieved by override control of the loop relating to the throughput valve downstream of the liquid storage tank (41D04/42D_02/43D02). If the anti-surge bypass opens, it must be ensured that the compressors are prevented from liquid entrance. *I.e.*, superheated vapor must be present downstream of the aircoolers 41E_01, 42E_01, and 43E_02. As this is not always fulfilled for the PC due to rather heavy mixed refrigerant, a temperature controller satisfies that a lower temperature limit after the 41E_01 is not declined.

When optimal operation of the MFC® process is considered, the question arises how to distribute the refrigeration/compressor load among the three cascaded cycles in order to achieve maximum LNG throughput. The answer to this question would be some kind of rule or feed-forward policy. For instance, Low *et al.* (1995/12/20) considered the similar question of how to transfer loads between drivers in adjacent refrigeration cycles and came up with a process, apparatus and control method. In this work, the question is restated. Rather than asking how to act on MVs in order to achieve the maximum LNG throughput, it is asked which CVs can be kept at fixed values in order to achieve the same objective. This question can be answered by applying the self-optimizing CSD framework.

4.7.1. Degree of freedom analysis

It is assumed that both, the LNG temperature controller and the controller for the temperature after the aircooler 41E_01, are closed and the respective MVs cannot be independent inputs. The degree of freedom analysis by Najim (1989, pp. 408-410) and Jensen and Skogestad (2007b, pp. 408-410) revealed $n_c + 2$ degrees of freedom for a simple mixed refrigerant cycle with variable active charge and fixed heat transfer area of the condenser and evaporator. As the MFC® process consists of three such cycles, a degree of freedom of $3\,n_c + 6$ is obtained. They are itemized as follows:

3×1 Compressor speeds n_r^{PC}, n_r^{LC} and n_r^{SC}. Note that in the single shaft configuration,

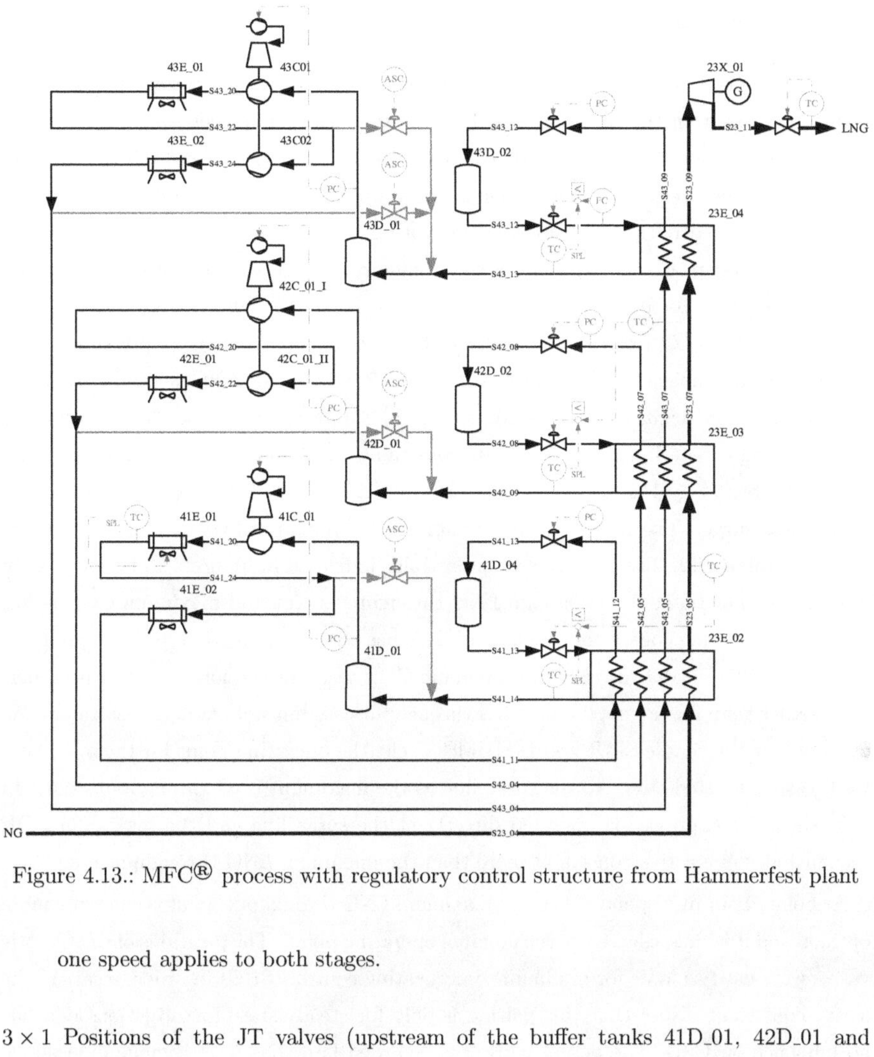

Figure 4.13.: MFC® process with regulatory control structure from Hammerfest plant

one speed applies to both stages.

3×1 Positions of the JT valves (upstream of the buffer tanks 41D_01, 42D_01 and 43D_01). They dominantly affect the discharge/suction pressure ratios, π_{PC}, π_{LC} and π_{SC}, and, as a result, the active charges in the cycles.

3×1 Positions of throughput valves located downstream of the buffer tanks. They dominantly affect mixed refrigerant flow rates. Instead of the valve positions, the flow rates of the mixed refrigerants, F_{PC}, F_{LC} and F_{SC}, are considered MVs. *I.e.*, it is supposed that flow controllers manipulate the valve positions.

$3 \times (n_{\mathbf{c}} - 1)$ Mixed refrigerant compositions

4.7.2. Model setup

The basis of the CSD was a design flowsheet of the Cigma study mentioned above. The transformation from the design into a dynamic simulation flowsheet was carried out as described in Section 4.2. As no partly condensation/separation of mixed refrigerant takes place within the cycles and their active charges are variable, the compositions and flow rates of the mixed refrigerants are independent on process conditions. Accordingly, the cycles could be modeled in partly closed configuration for steady-state investigations (see Section 2.5.1 for details). The COP and the LNG throughput were optimized by variation of MVs, *i.e.*, mixed refrigerants' flow rates and compression pressure ratios, subject to an ambient temperature of $T_{\mathrm{amb}} = 298.15\,\mathrm{K}$, an LNG setpoint temperature of $T_{\mathrm{LNG}}^{\mathrm{sp}} = 109.35\,\mathrm{K}$, compressor speeds $n_{\mathrm{r}}^{\mathrm{PC}} = n_{\mathrm{r}}^{\mathrm{LC}} = n_{\mathrm{r}}^{\mathrm{SC}} = 3000$, as well as the operating constraints. The operating constraints are limitations for suction pressure $\left(p_{\mathrm{suc}}^{k} \geq 2.0\,\mathrm{bar}\,\forall k \in \{\mathrm{PC},\mathrm{LC},\mathrm{SC}\}\right)$, superheating $\left(\Delta T_{\mathrm{SH}}^{k} \geq 10\,\mathrm{K}\,\forall k \in \{\mathrm{PC},\mathrm{LC},\mathrm{SC}\}\right)$ and compressor surge. The results for maximum LNG throughput and maximum COP are shown in Table 4.12. The objective values are also indicated in Figure 4.15 for the sake of illustration. For both, the maximum LNG throughput point and maximum COP point, all the superheating constraints are active. This has already been observed by Jensen and Skogestad (2006/07/09-13). For the maximum COP case, the suction pressures are generally smaller than for the maximum LNG throughput case but not at their constraints. As observed for the simple SMR and LIMUM® cycle, the operating point for the maximum COP case is located closer to the surge line as the maximum LNG throughput point. In fact, the LC compressor is operated directly at the surge line and the maximum COP case includes one active constraint more than the maximum LNG throughput case.

As pointed out in Appendix 4.A, the maximum LNG throughput point is generally more optimal and it is thus selected as the nominal operating point. The positions of the throttle valves were used as MVs for minimum superheating control (SHC) in order to satisfy the active constraint. Note that this pairing is only for steady-state investigations and has no influence on CSD. The actual pairing is regarded later based on dynamic measures. Ultimately, there are three MVs left for CSD. The feasibly region of the model was detected by parameter studies of the input variables. The resulting ranges are indicated in Figure 4.13. Instead of independent investigation of compressor speeds, they were synchronized, *i.e.*, $n_{\mathrm{r}} = n_{\mathrm{r}}^{\mathrm{PC}} = n_{\mathrm{r}}^{\mathrm{LC}} = n_{\mathrm{r}}^{\mathrm{SC}}$. Note that the speed range is fairly small due to the use of GE Frame 9 gas turbines as compressor drivers. Under variation of the ambient temperature, the MFC® process turned out to be more flexible (large feasibility) than both, the SMR and the LIMUM® cycle. However, it is more inflexible in terms of varying the LNG setpoint temperature. It was confirmed that minimum superheating is everywhere optimally active by running optimizations at various points close to the edge

	max COP	max F_{LNG}
COP (in %)	93.55	85.38
F_{LNG} (in mol/s)	20685.9	27410.7
F_{PC} (in mol/s)	25720.0	36172.1
F_{LC} (in mol/s)	17673.5	26464.2
F_{SC} (in mol/s)	8980.0	17572.3
π_{PC} (in bar)	4.2	3.6
π_{LC} (in bar)	7.9	6.2
π_{SC} (in bar)	21.5	13.4
$\Delta T_{\text{SH}}^{\text{PC}}$, $\Delta T_{\text{SH}}^{\text{LC}}$, $\Delta T_{\text{SH}}^{\text{SC}}$ (in K)	10.0	10.0
$p_{\text{suc}}^{\text{PC}}$ (in bar)	5.5	6.7
$p_{\text{suc}}^{\text{LC}}$ (in bar)	3.2	4.3
$p_{\text{suc}}^{\text{SC}}$ (in bar)	2.2	3.7
$\Delta F_{\text{surge}}^{\text{PC}}$ (in mol/s)	1084.0	6137.1
$\Delta F_{\text{surge}}^{\text{LC}}$ (in mol/s)	0.0	4514.1
$\Delta F_{\text{surge}}^{\text{SC}}$ (in mol/s)	893.9	4046.2

Table 4.12.: Nominal values at optimal operating points of the Cigma study

	Lower bound	Nominal point	Upper bound
F_{PC} (in mol/s)	30000 (−17.1 %)	36172.1	40000 (+10.6 %)
F_{LC} (in mol/s)	20000 (−24.4 %)	26464.2	30000 (+13.4 %)
F_{SC} (in mol/s)	12500 (−28.9 %)	17572.3	19000 (+8.1 %)
n_{r} (in RPM)	2880 (−4.0 %)	3000	3060 (+2.0 %)
$T_{\text{LNG}}^{\text{sp}}$ (in K)	109.0 (−0.3 %)	109.35	109.6 (+0.2 %)
T_{amb} (in K)	288.15 (−3.4 %)	298.15	308.15 (+3.4 %)

Table 4.13.: Perturbation range of the Cigma study

of the operating region.

4.7.3. CSD

The MV and DV vectors are respectively given by $\boldsymbol{u} = \begin{bmatrix} F_{\text{PC}} & F_{\text{LC}} & F_{\text{SC}} \end{bmatrix}^T$, *i.e.*, the PC, LC and SC flow, and $\boldsymbol{d} = \begin{bmatrix} T_{\text{amb}} & T_{\text{LNG}}^{\text{SP}} & n_{\text{r}} \end{bmatrix}^T$, *i.e.*, ambient temperature, LNG set point temperature and compressor speed. The PV vector $\boldsymbol{y}$ consists of 24 variables indicated in Table 4.14. The profit function is the LNG throughput, *i.e.*, $J = F_{\text{LNG}}$. According to these specifications, the steady-state I/O model and the Hessian were obtained at the nominal point. Note that the Hessian was calculated by permutation of MVs and DVs over the ranges given in Table 4.9 and using least squares fit of (3.3) on the resulting LNG throughput. The scaling matrix for the disturbances was set to $\boldsymbol{W_d} = \text{diag}\,(10\,\text{K}, 1\,\text{K}, 100\,\text{RPM})$. For the scaling matrix representing the implementation error, $\boldsymbol{W_{n^y}}$, it was assumed that 1 % flow uncertainty, 0.5 K absolute temperature uncertainty and 10 mbar absolute pressure uncertainty are present.

Stream	F	T	p
S41_11	1	5	19
S42_04	2	6	20
S43_04	3	7	21
S41_12		8	
S42_05		9	
S43_05		10	
S23_05	4	11	
S42_07		12	
S43_07		13	
S23_07		14	
S43_09		15	
S41_14		16	22
S42_09		17	23
S43_13		18	24

Table 4.14.: Measurement and their indices in $\boldsymbol{y}$ for the MFC® process

With the presented information, two a priori known control structures, MV0, CONV and JS06, were judged in terms of expected worst-case/average loss. MV0 refers to the structure in which F_{PC}, F_{LC} and F_{SC} remain unaffected. CONV relates to a control structure similar as indicated in Figure 4.13. In order to use the I/O model and Hessians, it must be assumed that minimum SHC is always active in each cycle instead of overriding the CV set $\{T_{\mathrm{S23\_05}}, T_{\mathrm{S43\_07}}, F_{\mathrm{SC}}\}$. The worst-case/average loss figures of the CONV structure are thus expected to underestimate the actual behavior. The JS06 structure refers to a structure similar to that proposed by Jensen and Skogestad (2006/07/09-13) based on heuristic considerations. The only modifications are that the loops which use compressor regulation were opened and the LNG temperature was controlled as indicated in Figure 4.13. The results of the worst-case/average loss and the measure $\Delta_{\mathrm{wc}}J$ defined in (4.4c) are presented in Table 4.15 for all considered structures. For better illustration, all values are related to the nominal throughput and are given in %. Both, the MV0 and the CONV structure, show fairly small loss figures in comparison to the JS06 structure. L_{wc} and $\Delta_{\mathrm{wc}}J$ of MV0 equal the measures $L_{\mathrm{wc},0}$ and $\Delta_{\mathrm{wc}}J_0$ as defined in (4.4a) and (4.4b), respectively. As $L_{\mathrm{wc},0}$ and $\Delta_{\mathrm{wc}}J_0$ are not too small and as $L_{\mathrm{wc},0} \ll \Delta_{\mathrm{wc}}J_0$ is not true, feedback control can be reasonably applied for optimization purposes.

The SEL structure in Table 4.15 relates to the PV selection structure with least worst-case loss and was obtained by applying the B3WC method. It shows only moderately smaller loss than MV0 and CONV. The worst-case/average loss could be further reduced by taking PV combinations into account. The COMB structure was obtained by using the MIAV method. It has the least average loss of all individually-sized control structures which satisfy the selection of temperatures and pure unit combination of maximum two

Name	Linear control structures	L_{av} (in %)	L_{wc} (in %)	$\Delta_{\text{wc}} J$ (in %)
MV0	$\{F_{\text{PC}}, F_{\text{LC}}, F_{\text{SC}}\}$	0.70	11.39	46.47
CONV	$\{p_{\text{S41\_12}}, p_{\text{S42\_07}}, p_{\text{S43\_09}}\}$	0.53	8.17	40.29
JS06	$\{p^{\text{PC}}_{\text{suc}}, p^{\text{LC}}_{\text{suc}}, p^{\text{SC}}_{\text{suc}}\}$	4.03	59.03	87.83
SEL	$\{T_{\text{S41\_11}}, F_{\text{LC}}, p_{\text{S43\_09}}\}$	0.28	3.87	38.82
COMB	$\left\{\begin{array}{l} F_{\text{PC}} - F_{\text{LC}}, \\ p_{\text{S42\_07}} - 6.87\, p_{\text{S42\_09}}, \\ p_{\text{S43\_09}} - 11.52\, p_{\text{S43\_13}} \end{array}\right\}$	3.1e−2	0.51	35.52
HHC	$\left\{\begin{array}{l} F_{\text{LNG}} - 1.65\, F_{\text{PC}}, \\ T_{\text{S23\_05}}, \\ F_{\text{LNG}} - 0.77\, F_{\text{SC}} \end{array}\right\}$	9.8e−2	1.60	36.74

Table 4.15.: Worst-case/average loss of control structures for the MFC® process

pressures and two flows.

If separation of HHCs from the natural gas takes place, the temperature of the stream S23_05 needs to be fixed. Suppose that the LC flow rate serves as an MV for a respective controller loop. Then the dimension of the CSD problem is reduced by one MV and the coefficient matrices of the I/O model and the Hessian need to be transformed. This can take place as indicated in Appendix 4.C. With the transformed matrices, the HHC structure was calculated. It was obtained by the use of the MIAV method and has the least average loss among all individually-sized structures which satisfy the selection of temperatures and pressures as well as pure unit combinations of a maximum of two flow rates.

For all the so far considered structures in Table 4.15, only the COMB and the HHC structure satisfy the relationship $L_{\text{wc}} \ll \Delta_{\text{wc}} J$. Accordingly, if loss figures represent the nonlinear behavior sufficiently well, the optimization by feedback control using either of both structures achieves good performance and optimization of setpoints by RTO is expendable.

4.7.4. Nonlinear verification

The a priori known structures and newly developed structures were applied to the nonlinear OPTISM model for the sake of verification. The results are presented in Figure 4.14. The charts in the first row represent the LNG throughput vs. the three DVs. The R/O case represents the best achievable behavior (upper bound for feedback control structures) and was obtained via optimization of the OPTISIM® model. As the resolution in the first row of charts is rather poor, the deviation between the R/O curve and all other curves is respectively plotted in the diagrams in the second row. Expectedly, the maximum achievable LNG throughput (R/O) is monotonically decreasing with rising

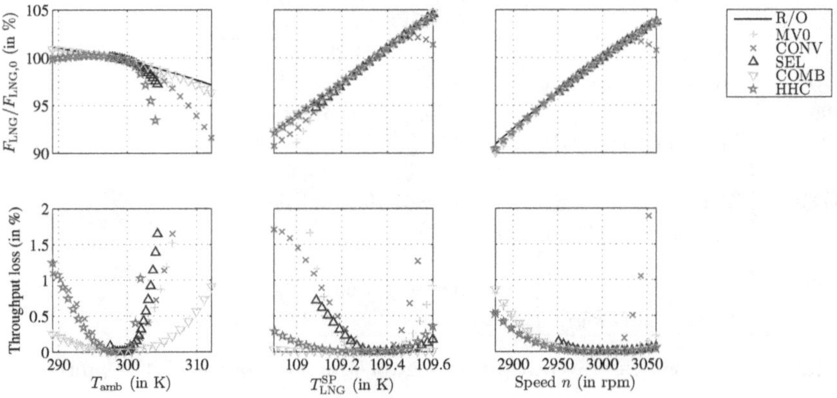

Figure 4.14.: Results of nonlinear analysis of control structures for the MFC® process

ambient temperature, falling LNG setpoint temperature and falling compressor speed. An ending curve indicates the edge of the feasibility region except for high ambient temperatures. There, ending curves are due to the fact that the surge line of one of the compressors is reached. The curves are discontinued at the surge line in order to provide better comparison between the curves and the worst-case loss figures.

It was observed that the worst-case loss figures of the MV0, CONV and JS06 structure are in good agreement with their nonlinear behavior. Due to considerably small feasibility region in all DVs, the JS06 structure is not included in the chart. It must be stressed that due to the suction pressure control, the feasibility of the CONV structure is actually smaller than indicated. The SEL structure is better than the MV0 and CONV except at high ambient temperatures. Therefore, the worst-case loss of the SEL structure underestimates its global behavior, whereas the average loss of the MV0 structure may agree well. Besides, the SEL structure shows slightly reduced feasibility region. The advance of the COMB over the SEL structure in terms of worst-case/average loss can be verified by its nonlinear behavior. It has the further advantage over so far considered structures that the surge line is not reached within the observed operating range. The nonlinear behavior of the COMB structure is excellent and confirms that setpoint optimization by RTO may be expendable if this structure is used.

4.8. Conclusions

A general CSD procedure for industrial processes was suggested and applied to an evaporation process and three LNG liquefaction processes. All types of control structures were taken into account and the advantages of common-sized and individually-sized control structures were pointed out. For LNG liquefaction processes, the best control structures in terms of least local worst-case/average loss were tested by nonlinear parameter studies. Not all promising control structures were positively verified by investigation of the nonlinear behavior. It is important to note that the newly developed control structures have the advantage that they are deviations from the conventional structures and can thus be transferred from one into the other during operation in a loop-by-loop fashion.

One important observation was that holding the natural gas temperature after the precooler fixed leads to an undesirable nonlinear behavior for both, the LIMUM® cycle and the MFC® process. What both processes have in common and distinguish themselves from the simple SMR cycle is that they loose degrees of freedom for optimization due to the optimality constraint of minimal superheating. The conclusion can be drawn that the combination of both CVs, natural gas temperature after the precooler and superheating of the mixed refrigerant, are somewhat counterproductive in maximizing the LNG throughput. This gives rise to the following design recommendation.

Rule 4.6. *If minimum superheating is an optimally active constraint in an LNG liquefaction process, then it is recommended to perform the potential HHC separation upstream of the cycle as both, fixing the superheating and the natural gas temperature after the precooler, leads to rather poor self-optimizing control behavior.*

For the development of self-optimizing control structures for LNG liquefaction processes, the manipulation of mixed refrigerant composition for optimization purposes was disregarded. This is due to the fact that dosing and venting is a comparably expensive manipulation. However, it may be reasonably applied for compensating seasonal fluctuations. It is expected that optimizing the LNG throughput by variation of the mixed refrigerant composition is a convex problem and may be solved by the use of online optimization techniques such as RTO. Considering the steady-state modeling issues discussed in Section 2.5.1, this is expected to be a rather challenging task. By best knowledge of the author, this has so far neither been investigated nor implemented and may be a subject for future work.

4.A. Objective function

There are two candidate objectives, the coefficient of performance

$$\mathrm{COP} = \frac{F_{\mathrm{LNG}}\,(h_{\mathrm{NG}} - h_{\mathrm{LNG}})}{W_{\mathrm{s}}} \tag{4.5}$$

as defined in the textbook of Haywood (1980, p. 75) and the LNG throughput F_{LNG}. Here, it is discussed which of both is the most appropriate objective in terms of generality.

A decision can be drawn by considering the economic profit function

$$J = c_1\, F_{\mathrm{LNG}} - c_2\, W_{\mathrm{s}} \tag{4.6}$$

which states the net income as the difference of product sales and shaft power cost, *i.e.*, c_1 indicates the sales price of the LNG in \$/(kmol/s) and c_2 indicates the power generation cost in \$/MW. Eliminating the shaft power W_{s} using definition of the COP (4.5) yields

$$\frac{J}{c_1} = F_{\mathrm{LNG}} \left(1 - \frac{\gamma}{\mathrm{COP}}\right) \tag{4.7}$$

where $\gamma = c_2/c_1\,(h_{\mathrm{NG}} - h_{\mathrm{LNG}})$ is an indicator for project and market conditions. In order to come to a general conclusion it is convenient to determine the lower and upper bound of γ. Suppose, waste gas is used as fuel for the generation of shaft power. This then corresponds to zero cost and the lower bound $\gamma \geq 0$. The upper bound is obtained by using the most expensive fuel available onsite which is of course the LNG itself. The cost of shaft power generation using a gas turbine is then given by $c_2 = c_1/\,(\mathrm{LHV}\,\eta_{\mathrm{GT}})$ where LHV indicates the lower heating value of the (vaporized) LNG and η_{GT} the gas turbine efficiency. Based on a heating value of 40.9 MJ/m^3 for a typical natural gas at standard conditions (after Katz and Lee, 1990, p. 102), a relatively inefficient gas turbine, *i.e.*, $\eta_{\mathrm{GT}} = 0.2$ (after Soares, 2002, p. T57) and a enthalpy difference $h_{\mathrm{NG}} - h_{\mathrm{LNG}}$ for the Mossel Bay plant of 15 MJ/kmol, the upper bound $\gamma \leq 0.08$ can be calculated. Using the lower and upper bound for γ, two contour plots shown in Figure 4.15 have been created. They represent the economic profit function (4.7) versus LNG throughput and COP. As additional information, the maximum throughput and maximum COP operating points of three LNG liquefaction processes of different types and scales are included in the graph. It is obvious that for any value of γ and for all cycles, the LNG throughput is the better representation of the profit function. This serves as a proof for selecting the LNG throughput as a general replacement of the profit function (4.7). Note that for $\gamma = 0$, the profit function is in fact the scaled LNG throughput.

Remark 4.7. Considering a proprietary SMR cycle and the same profit function (4.6), Michelsen *et al.* (2010) came to the conclusion that the optimum operating point is given at maximum cooling capacity (maximum throughput and maximum compressor speed).

Remark 4.8. Aske *et al.* (2008) mentioned earlier that for some plants and market conditions (*e.g.*, large sales price) the economic optimum and the maximum throughput point almost agree. Considering the PRICO cycle, Jensen and Skogestad (2009a) came to similar conclusions.

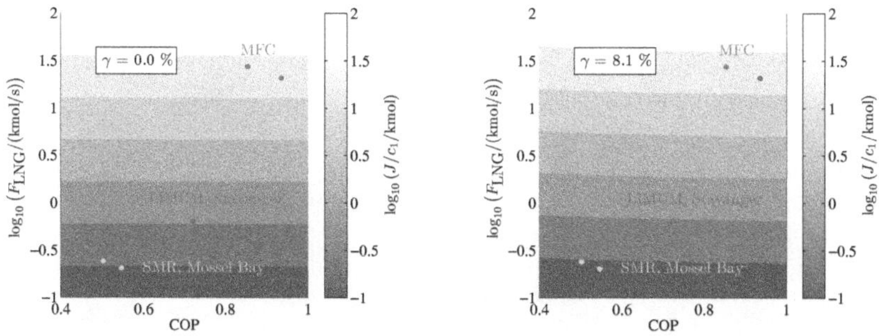

(a) Zero cost of power generation (b) Maximum cost of power generation

Figure 4.15.: Consideration of objective functions for operation

4.B. Evaporation process model

The DAE of the evaporation process is given by

$$\frac{\mathrm{d}}{\mathrm{d}t} L_2 = \frac{1}{20.0} (F_1 - F_4 - F_2) \tag{4.8a}$$

$$\frac{\mathrm{d}}{\mathrm{d}t} X_2 = \frac{1}{20.0} (F_1 X_1 - F_2 X_2) \tag{4.8b}$$

$$\frac{\mathrm{d}}{\mathrm{d}t} p_2 = \frac{1}{4.0} (F_4 - F_5) \tag{4.8c}$$

$$T_2 = 0.5616\, p_2 + 0.3126\, X_2 + 48.43 \tag{4.8d}$$

$$T_3 = 0.507\, p_2 + 55.0 \tag{4.8e}$$

$$F_4 = \frac{1}{38.5} (Q_{100} - 0.07\, F_1\, (T_2 - T_1)) \tag{4.8f}$$

$$T_{100} = 0.1538\, p_{100} + 90.0 \tag{4.8g}$$

$$Q_{100} = 0.16\, (F_1 + F_3)\, (T_{100} - T_2) \tag{4.8h}$$

$$F_{100} = \frac{1}{36.6} Q_{100} \tag{4.8i}$$

$$Q_{200} = 0.9576\, F_{200} \frac{T_3 - T_{200}}{0.14\, F_{200} + 6.84} \tag{4.8j}$$

$$T_{201} = T_{200} + 13.68 \frac{T_3 - T_{200}}{0.14\, F_{200} + 6.84} \tag{4.8k}$$

$$F_5 = \frac{1}{38.5} Q_{200}. \tag{4.8l}$$

The operational constraints read

$$
\begin{aligned}
35\,\% + 0.5\,\% \leq\ & X_{002} & & \quad (4.9a)\\
40\,\text{kPa} \leq\ & p_{002} & \leq 80\,\text{kPa} & \quad (4.9b)\\
& p_{100} & \leq 400\,\text{kPa} & \quad (4.9c)\\
0\,\text{kg/min} \leq\ & F_{200} & \leq 400\,\text{kg/min} & \quad (4.9d)\\
0\,\text{kg/min} \leq\ & F_{001} & \leq 20\,\text{kg/min} & \quad (4.9e)\\
0\,\text{kg/min} \leq\ & F_{003} & \leq 100\,\text{kg/min}. & \quad (4.9f)
\end{aligned}
$$

4.C. Model reduction

Suppose that the subset $\mathcal{U}$ among the MV set $\mathcal{W} \supset \mathcal{U}$ is spend in order to fix additional CVs represented by $\boldsymbol{H}_{\mathcal{U}} \in \mathbb{R}^{|\mathcal{U}| \times n_y}$. Then the coefficient matrices of the I/O model and the Hessians need to be transformed in order to solve the CSD problem for the remaining MV set $\mathcal{V} = \mathcal{W} \backslash \mathcal{U}$. Splitting $\boldsymbol{u}$ into $\boldsymbol{u}_{\mathcal{U}}$ and $\boldsymbol{u}_{\mathcal{V}}$, (3.2) can be written as

$$
\boldsymbol{y} = \boldsymbol{G}^{\boldsymbol{y}}_{\boldsymbol{u}_{\mathcal{U}}} \boldsymbol{u}_{\mathcal{U}} + \boldsymbol{G}^{\boldsymbol{y}}_{\boldsymbol{u}_{\mathcal{V}}} \boldsymbol{u}_{\mathcal{V}} + \boldsymbol{G}^{\boldsymbol{y}}_{\boldsymbol{d}} \boldsymbol{d} + \boldsymbol{n}^{\boldsymbol{y}}. \tag{4.10}
$$

From

$$
\boldsymbol{H}_{\mathcal{U}} \boldsymbol{y} = \boldsymbol{c} \overset{!}{=} \boldsymbol{c}_{\mathrm{s}} = \boldsymbol{0}
$$

and (4.10) one obtains

$$
\begin{aligned}
\boldsymbol{u}_{\mathcal{U}} = & \underbrace{\left(-\boldsymbol{H}_{\mathcal{U}} \boldsymbol{G}^{\boldsymbol{y}}_{\boldsymbol{u}_{\mathcal{U}}}\right)^{-1} \boldsymbol{H}_{\mathcal{U}} \boldsymbol{G}^{\boldsymbol{y}}_{\boldsymbol{u}_{\mathcal{V}}}}_{=\boldsymbol{G}^{\boldsymbol{u}_{\mathcal{U}}}_{\boldsymbol{u}_{\mathcal{V}}}} \boldsymbol{u}_{\mathcal{V}} + \underbrace{\left(-\boldsymbol{H}_{\mathcal{U}} \boldsymbol{G}^{\boldsymbol{y}}_{\boldsymbol{u}_{\mathcal{U}}}\right)^{-1} \boldsymbol{H}_{\mathcal{U}} \boldsymbol{G}^{\boldsymbol{y}}_{\boldsymbol{d}}}_{=\boldsymbol{G}^{\boldsymbol{u}_{\mathcal{U}}}_{\boldsymbol{d}}} \boldsymbol{d}\\
& + \underbrace{\left(-\boldsymbol{H}_{\mathcal{U}} \boldsymbol{G}^{\boldsymbol{y}}_{\boldsymbol{u}_{\mathcal{U}}}\right)^{-1} \boldsymbol{H}_{\mathcal{U}}}_{=\boldsymbol{G}^{\boldsymbol{u}_{\mathcal{U}}}_{\boldsymbol{n}^{\boldsymbol{y}}}} \boldsymbol{n}^{\boldsymbol{y}}.
\end{aligned}
$$

Inserting this result into (4.10), the coefficient matrices and noise of the reduced model read

$$
\begin{aligned}
\left.\boldsymbol{G}^{\boldsymbol{y}}_{\boldsymbol{u}_{\mathcal{V}}}\right|_{\text{red}} &= \boldsymbol{G}^{\boldsymbol{y}}_{\boldsymbol{u}_{\mathcal{U}}} \boldsymbol{G}^{\boldsymbol{u}_{\mathcal{U}}}_{\boldsymbol{u}_{\mathcal{V}}} + \boldsymbol{G}^{\boldsymbol{y}}_{\boldsymbol{u}_{\mathcal{V}}}\\
\left.\boldsymbol{G}^{\boldsymbol{y}}_{\boldsymbol{d}}\right|_{\text{red}} &= \boldsymbol{G}^{\boldsymbol{y}}_{\boldsymbol{u}_{\mathcal{U}}} \boldsymbol{G}^{\boldsymbol{u}_{\mathcal{U}}}_{\boldsymbol{d}} + \boldsymbol{G}^{\boldsymbol{y}}_{\boldsymbol{d}}\\
\left.\boldsymbol{n}^{\boldsymbol{y}}\right|_{\text{red}} &= \boldsymbol{G}^{\boldsymbol{y}}_{\boldsymbol{u}_{\mathcal{U}}} \boldsymbol{G}^{\boldsymbol{u}_{\mathcal{U}}}_{\boldsymbol{n}^{\boldsymbol{y}}} \boldsymbol{n}^{\boldsymbol{y}} + \boldsymbol{n}^{\boldsymbol{y}}.
\end{aligned}
$$

Similarly, the reduced Hessians

$$J_{u_\mathcal{V} u_\mathcal{V}}\big|_{\text{red}} = \left(G_{u_\mathcal{V}}^{u_\mathcal{U}}\right)^T J_{u_\mathcal{U} u_\mathcal{U}} G_{u_\mathcal{V}}^{u_\mathcal{U}} + J_{u_\mathcal{V} u_\mathcal{V}}$$
$$J_{u_\mathcal{V} d}\big|_{\text{red}} = \left(G_{u_\mathcal{V}}^{u_\mathcal{U}}\right)^T J_{u_\mathcal{U} d} + J_{u_\mathcal{V} d}$$

can be obtained.

Bibliography

E04XAF: NAG Fortran library routine document. In *NAG Fortran library manual: Mark 21*. 2006b. ISBN 1852062045.

M. Abramowitz and I. A. Stegun. *Handbook of mathematical functions: With formulas, graphs, and mathematical tables*. Dover books on intermediate and advanced mathematics. Dover Publishing, New York, New York, 1970. ISBN 0486612724.

R. Agustriyanto and J. Zhang. Self optimizing control of an evaporation process under noisy measurements. International control conference, Glasgow, Scotland, 2006/08/30-09/01.

E. M. B. Aske, S. Strand, and S. Skogestad. Coordinator MPC for maximizing plant throughput. *Computers and chemical engineering*, 32(1-2):195–204, 2008.

M. Baldea, A. C. B. Araujo, S. Skogestad, and P. Daoutidis. Dynamic considerations in the synthesis of self-optimizing control structures. *American institute of chemical engineers journal*, 54(7):1830–1841, 2008.

H. C. Bauer. Best available techniques for the Hammerfest Snøhvit LNG project. *Scandinavian oil-gas magazine*, (11), 2005.

E. Berger, W. Förg, R. S. Heiersted, and P. Paurola. Das Snøhvit-Projekt: Der Mixed Fluid Cascade (MFC(R)) Prozess für die erste europäische LNG-Baseload-Anlage. *Linde technology*, (1):12–23, 2003a.

E. Berger, X. Dong, J. G. Qiang, A. Meffert, and L. Atzinger. LNG baseload plant in Xinjiang, China: Commercialisation of remote gas resources for an eco-responsible future. World gas conference, Tokyo, Japan, 2003/06/01-05.

Y. Cao. Constrained self-optimizaing control via differentiation. International symposium on advanced control of chemical processes, Hong Kong, China, 2004/01/11-14.

J. E. Dennis and R. B. Schnabel. *Numerical methods for unconstrained optimization and nonlinear equations.* Prentice-Hall series in computational mathematics. Prentice-Hall, Englewood Cliffs, New Jersey, 1983. ISBN 0136272169.

J. J. Downs and S. Skogestad. An industrial and academic perspective on plantwide control. In S. Engell and Y. Arkun, editors, *ADCHEM 2009: Preprints of IFAC symposium on advanced control of chemical processes: July 12-15, 2009, Koç University, Istanbul, Turkey*, volume 1, pages 119–130. 2009.

E. Eich-Soellner, P. Lory, P. S. Burr, and A. Kröner. Stationary and dynamical flowsheeting in the chemical industry. *Surveys on mathematics for industry*, 7:1–28, 1997.

S. Engell. Feedback control for optimal process operation. *Journal of process control*, 17 (3):203–219, 2007.

S. Engell, T. Scharf, and M. Völker. A methodology for control structure selection based on rigorous process models. IFAC world congress, Prague, Czech Republic, 2005/07/04-08.

W. Förg, R. Stockmann, M. Bölt, M. Steinbauer, C. Pfeiffer, P. Paurola, A. O. Fredheim, and O. Sorensen. Method for liquefying a stream rich in hydrocarbons, 2000/03/24.

P. E. Gill, W. Murray, M. A. Saunders, and M. H. Wright. Computing forward-difference intervals for numerical optimization. *SIAM journal on scientific and statistical computing*, 4:310–321, 1981.

I. J. Halvorsen, S. Skogestad, J. C. Marud, and V. Alstad. Optimal selection of controlled variables. *Industrial and engineering chemistry research*, 42:3273–3284, 2003.

R. W. Haywood. *Analysis of engineering cycles.* Pergamon international library of science, technology, engineering and social studies. Pergamon Press, Oxford, UK, 3rd ed. edition, 1980. ISBN 0080254403.

J. A. Heath, I. K. Kookos, and J. D. Perkins. Process control structure selection based on economics. *American institute of chemical engineers journal*, 46(10):1998–2016, 2000.

R. S. Heiersted. Snøhvit LNG project: Concept selection for Hammerfest LNG plant. Gastech, Doha, Qatar, 2002/10/13-16.

E. S. Hori and S. Skogestad. Selection of controlled variables: Maximum gain rule and combination of measurements. *Industrial and engineering chemistry research*, 47(23): 9465–9471, 2008.

E. S. Hori, S. Skogestad, and M. A. Al-Arfaj. Self-optimizing control configurations for two-product distillation columns. Distillation and absorption conference, London, UK, 2006/09/04-06.

J. E. P. Jäschke and S. Skogestad. Optimally invariant variable combinations for nonlinear systems. In S. Engell and Y. Arkun, editors, *ADCHEM 2009: Preprints of IFAC symposium on advanced control of chemical processes: July 12-15, 2009, Koç University, Istanbul, Turkey*, volume 2, pages 551–556. 2009.

J. E. P. Jäschke, H. Smedsrud, S. Skogestad, and H. Manum. Optimal operation of a waste incineration plant for district heating. American control conference, Hyatt Regency Riverfront, St. Louis, Missouri, 2009/06/10-12.

J. B. Jensen. Control and optimal operation of simple heat pump cycles. volume 20 of *European Symposium on Computer Aided Process Engineering*, Barcelona, Spain, 2005/05/29-06/01.

J. B. Jensen and S. Skogestad. Optimal operation of a mixed fluid cascade LNG plant. Symposium on Process Systems Engineering/European Symposium on Computer Aided Process Engineering, Garmisch-Partenkirchen, Germany, 2006/07/09-13.

J. B. Jensen and S. Skogestad. Optimal operation of simple refrigeration cycles: Part II: Selection of controlled variables. *Computers and chemical engineering*, 31(12):1590–1601, 2007a.

J. B. Jensen and S. Skogestad. Optimal operation of simple refrigeration cycles: Part I: Degrees of freedom and optimality of sub-cooling. *Computers and chemical engineering*, 31(5-6):712–721, 2007b.

J. B. Jensen and S. Skogestad. Single-cycle mixed-fluid LNG process: Part II: Optimal operation. In H. E. Alfadala, G. V. R. Reklaitis, and M. M. El-Halwagi, editors, *Proceedings of the 1st annual gas processing symposium: 10 - 12 January 2009, Doha, Qatar*, pages 221–226. Elsevier, 2009a. ISBN 9780444532923.

J. B. Jensen and S. Skogestad. Single-cycle mixed-fluid LNG process: Part I: Optimal design. In H. E. Alfadala, G. V. R. Reklaitis, and M. M. El-Halwagi, editors, *Proceedings of the 1st annual gas processing symposium: 10 - 12 January 2009, Doha, Qatar*, pages 213–220. Elsevier, 2009b. ISBN 9780444532923.

J. B. Jensen and S. Skogestad. Steady-state operational degrees of freedom with application to refrigeration cycles. *Industrial and engineering chemistry research*, 48(14): 6652–6659, 2009c.

V. Kariwala and Y. Cao. Bidirectional branch and bound for controlled variable selection: Part II. Exact local method for self-optimizing control. *Computers and chemical engineering*, 2009.

V. Kariwala and Y. Cao. Bidirectional branch and bound for controlled variable selection: Part III. Local average loss minimization. *IEEE transactions on industrial informatics*, 2010a.

V. Kariwala, Y. Cao, and S. Janardhanan. Local self-optimizing control with average loss minimization. *Industrial and engineering chemistry research*, 47(4):1150–1158, 2008.

D. L. V. Katz and R. L. Lee. *Natural gas engineering: Production and storage.* McGraw-Hill chemical engineering series. McGraw-Hill, New York, New York, 1990. ISBN 0070333521.

T. Larsson, K. Hestetun, E. Hovland, and S. Skogestad. Self-optimizing control of a large-scale plant: The Tennessee Eastman process. *Industrial and engineering chemistry research*, 40(22):4889–4901, 2001.

V. Lersbamrungsuk. *Development of control structure design and structural controllability for heat exchanger networks: Ph.D. thesis.* Ph.D. thesis, Kasetsart University, Bangkok, Thailand, 2008.

W. R. Low, D. L. Andress, and C. G. Houser. Method of load distribution in a cascaded refrigeration process, 1995/12/20.

K. H. Lüdtke. *Process centrifugal compressors: Basics, function, operation, design, application.* Springer, Berlin, Germany, 2004. ISBN 3540404279.

J. A. Mandler and P. A. Brochu. Controllability analysis of the LNG process. AIChE annual meeting, Los Angeles, Californien, Nov 1997.

J. A. Mandler, P. A. Brochu, and J. R. Hamilton. Method and apparatus for regulatory control of production and temperature in a mixed refrigerant liquefied natural gas facility, 1997/07/24.

J. A. Mandler, P. A. Brochu, J. Fotopoulos, and L. Kalra. New control strategies for the LNG process. International conference & exhibition on Liquefied Natural Gas, Perth, Australia, May 1998.

G. K. McMillan. *Centrifugal and axial compressor control*, volume Student text. Instrument Society of America, Research Triangle Park, North Carolina, 1983. ISBN 0876647441.

F. A. Michelsen, B. F. Lund, and I. J. Halvorsen. Selection of optimal, controlled variables for the TEALARC LNG process. *Industrial and engineering chemistry*, 49(18):8624–8632, 2010.

K. Najim. *Process modeling and control in chemical engineering*, volume 38 of *Chemical industries*. Dekker, New York, New York, 1989. ISBN 0824782046.

R. B. Newell and P. L. Lee. *Applied process control: A case study*. Prentice-Hall, New York, New York, 1989. ISBN 0130409405.

C. L. Newton. Automated control system for a multicomponent refrigeration system, 1986/07/10.

J. Oldenburg, H.-J. Pallasch, C. Carroll, V. Hagenmeyer, S. Arora, K. Jacobsen, J. Birk, A. Polt, and P. van den Abeel. Decision support for control structure selection during plant design. Symposium on Process Systems Engineering/European Symposium on Computer Aided Process Engineering, Lyon, France, 2008/06/01-04.

H. Reithmeier, M. Steinbauer, R. Stockmann, A. O. Fredheim, O. Jørstad, and P. Paurola. Industrial scale testing of a spiral wound heat exchanger. LNG conference, Doha, Qatar, 2004/04/21-24.

A. Singh and M. Hovd. Mathematical modeling and control of a simple liquefied natural gas process. Scandinavian simulation society, Helsinki, Finland, 2006/09/28-29.

S. Skogestad. Control structure design for complete chemical plants. *Computers and chemical engineering*, 28(1-2):219–234, 2004a.

C. M. Soares. *Process engineering equipment handbook*. McGraw-Hill handbooks. McGraw-Hill, New York, New York, 2002. ISBN 007059614X.

R. Stockmann, M. Bölt, M. Steinbauer, C. Pfeiffer, P. Paurola, W. Förg, A. O. Fredheim, and O. Sorensen. Process for liquefying a hydrocarbon-rich stream, 1997/05/28.

E. Voskresenskaya. Potential application of floating LNG. Global conference on renewables and energy efficiency for desert regions, Amman, Jordan, 2009/03/31-04/02.

5. Operability analysis of LNG liquefaction processes

In Chapter 4, control structures for LNG liquefaction processes have been proposed and judged with respect to steady-state economic measures. It is important to note that in this context the term control structure refers to a set of control variables (CVs) which are combinations of measurable process variables (PVs). Another design issue which is the subject of this chapter is the mapping between manipulated variables (MVs) and CVs for decentralized control. In order to distinguish it from the term control structure, the term control strategy is introduced. It refers to the entity of the three design aspects, selection of CV set, selection of MV set and the pairing between CVs and MVs. The dynamic behavior of a decentralized controlled process depends on theses three design aspects. This section is concerned with comparison of publicly know and newly derived control structures in terms of dynamic measures. This is referred to as operability analysis according to the definition of Wolff (1994, p. 4).

5.1. Related work

Several Ph.D. students such as Wolff (1994); Havre (Jan 1998) and (Lausch, 1999) dedicated their work to the operability analyses of chemical processes. That operability analysis is not only an academic field is proven by various industrial applications. For instance, (Zapp, 1994) describes how inherently bad or even instable pairing combinations can be recognized at an early stage of a plant construction project of an air separation unit. He compared competitive pairing structures with frequency dependent measures and proved his result by the use dynamic simulation. Operability analysis and control design of LNG liquefaction processes have been considered by Mandler and Brochu (Nov 1997); Mandler *et al.* (May 1998) and Singh and Hovd (2006/09/28-29). Mandler and Brochu (Nov 1997); Mandler *et al.* (May 1998) proposed control strategies for the C3MR process and came to the conclusion that it is beneficial to rethink the traditional way of controlling the LNG product temperature by manipulation of the LNG flow rate. Singh and Hovd (2006/09/28-29) considered the PRICO cycle (Price and Mortko, 1996/12/3-6) whose topology is similar to that of the LIMUM® cycle.

Dynamical simulations of LNG liquefaction processes have been considered by various authors. The early work of Melaaen (Oct 1994) demonstrated modeling, dynamic simulation and control of an SWHE network. Zaïm (Mar 2002) developed a rigorous dynamic model of a C3MR process for the purpose of predicting the productivity of a compound of trains. His idea was to minimize the boil-off subject to ensured delivery to randomly arriving LNG carriers. Hammer developed a rigorous model of the MFC® process in the configuration of the Hammerfest plant introduced in Section 4.7. He focused rather on thermodynamic, mathematical and computational aspects than on system analysis and process control. Note that in Section 5.5, the MFC® process in a slightly different configuration than that in the work of Hammer is taken into account. In a further work of Okasinski and Schenk (2007/04/ 24-27), the C3MR and AP-X™ processes were investigated by dynamical simulations. Two events were studied, a propane cycle discharge event due to loss of cooling water and the transition from AP-X™ to C3MR by taking the subcooling (N2) cycle out of operation.

5.2. Motivation

LNG liquefaction processes are dedicated to one or two tasks. The mandatory task is the precooling, liquefaction and subcooling of the natural gas to a predefined temperature. The optional task is keeping the temperature of the precooled natural gas fixed due to the separation of HHCs. Process control of LNG liquefaction processes is concerned with the reliable and tight accomplishment of these tasks under the influence of disturbances. As this problem is somewhat difficult to capture, the operational targets and related operability measures are categorized and listed in Table 5.1. Product quantity has been comprehensively considered in Chapter 3. Product quality refers to keeping the specifications of the products within their tolerances, *i.e.*, temperature and composition of the LNG, despite disturbances acting on the process. This is known as setpoint tracking and disturbance rejection and measures thereof are the PRGA $\boldsymbol{\Gamma}$ and the CLDG $\tilde{\boldsymbol{G}}_{\boldsymbol{d}}$, respectively. The term reliability refers to stable operation despite of varying disturbances and controller failure, *etc.* Related measures are the RGA $\boldsymbol{\Lambda}$, the NI, the PRGA and the CLDG. Versatility relates to the ease of changing operating regions. For instance, a plant which has a small feasible operating region (MSV) is not considered versatile[1]. Further, a plant which needs long settlement times when changing between operating points (RHPZ) or whose gains depend strongly on the directions (CN, DCN) is neither considered versatile. A brief survey of operabilty measures is given in Appendix 5.A.

[1]This is sometimes referred to as flexibility or static resiliency (Lima *et al.*, 2010).

Category	Operational targets	Measures[a]
Product quantity	Maximizing LNG throughput	L_{wc}, L_{av}
Product quality	Keeping LNG temperature in tight bounds	$\mathbf{\Gamma}$, $\tilde{\boldsymbol{G}}_d$
	Keeping natural gas precooling temperature in tight bounds	$\mathbf{\Gamma}$, $\tilde{\boldsymbol{G}}_d$
Reliability	Ensuring superheated refrigerant at compressor inlet	$\mathbf{\Gamma}$, $\tilde{\boldsymbol{G}}_d$
	Avoiding too low suction pressure[b]	$\mathbf{\Gamma}$, $\tilde{\boldsymbol{G}}_d$
	Preventing from driver overload[b]	$\mathbf{\Gamma}$, $\tilde{\boldsymbol{G}}_d$
	Ensuring stability	NI
	Ensuring integrity of control loops	$\mathbf{\Lambda}$
Versatility	Providing fast transition between steady-state modes	RHPZ
	Avoiding active limitations of MVs	MSV
	Providing independence on I/O directions	CN, DCN

[a]Readers not familiar with these operability measures are referred to Appendix 5.A.
[b]Out of scope as compressor regulation is disregarded.

Table 5.1.: Operational targets

As introduced in Section 4.4.1, the major DVs acting on LNG liquefaction processes are ambient temperature, LNG setpoint temperature and compressor regulation. A further DV is the composition of the feed gas which happens to change considerably due to reliquefaction of boil-off gas generated during ship loading. Reliquefaction of boil-off gas becomes popular as more and more operating companies follow a zero flaring policy. During ship loading, the ship return gas needs to be recompressed for feeding it into the natural gas upstream of the LNG liquefaction process. This disturbance in natural gas composition is discontinuous and acts on the process only during a limited time frame. The effect of reliquefaction of boil-off gas is taken into account in this work. It is represented by the factor $\mathrm{S} \in \{0, 1\}$, whereas a value of 1 refers to lighter natural gas, *i.e.*, to the case of reliquefaction of boil-off gas.

5.3. General considerations

Fundamental limitations on the bandwidth of controlled systems are introduced by RHPZs close to the imaginary axis and by time delays. Suppose that none of both is present in a plant model. The question arises whether the plant is then really limitation free? In most cases, the answer is no due to model-plant mismatch which usually increases with frequency. For instance, shock waves in piping/equipment which act on a fairly short time scale are generally not modeled (the momentum balance is often assumed quasi-stationary) for the sake of simplicity and computational load reduction. Accordingly,

Type	Var.	Description	Nom. point	Span
MVs	FP	LNG flow rate (in mol/s)	644.1	60
	F1	MHMR flow rate (in mol/s)	410.9	40
	F2	LMR flow rate (in mol/s)	621.7	60
DVs	Z	IGV angles (in °)	20.0	40.0
	TA	Ambient temperature (in K)	296.0	10.0
	S	Ship return gas $\in [0, 1]$	0.0	1.0
Cand. CVs	TP	LNG temperature (in K)	111.66	2.0
	TN	Precooled natural gas temp. (in K)	239.6	5.0
	SH	Degree of superheating (in K)	10.0	5.0
	FS	LMR plus HMR flow rate (in mol/s)	1434.6	140
	FR	LMR & HMR flow rate relation	1500.1	150

Table 5.2.: I/O variables of the state-space model of the LIMUM® cycle

although no fundamental limitations can be found in the plant model, it may be convenient to impose a lumped limitation in order to account for unmodeled effects.

In this work, the most dominant modeling inaccuracy is probably due to disregarding piping in the models. Further source of error may be SWHE model simplifications whose effects on dynamical model accuracy are discussed in Section 2.4.5.2. A lumped time delay of 20 s is assumed for all LNG liquefaction processes considered here. This relates to an upper limit on the achievable bandwidth of $f_b < 0.008\,\text{Hz}$ ($\omega_b < 0.05\,\text{rad/s}$). It must be stressed that this bandwidth limitation includes other unmodeled effects such as valve dynamics *etc.*, predominantly present at small time scales. This practice is in agreement with that of Singh and Hovd (2006/09/28-29) and leads to more general conclusions. For instance, valve dynamics depend on the particular fitting/driver architecture whose dynamics correlate with the time scale of the plant.

All investigations in frequency domain are performed within the interval $\left[10^{-6}, 10^{-2}\right]$ Hz. The upper bound relates to the bandwidth discussed above. LNG plant dynamics do not occur below the lower limit. Besides, every process simulator has limited computational accuracy which limits the frequency band likewise. For instance, an integrator in a complex OPTISIM® flowsheet generally relates to a pole very close to but not exactly at the origin of the complex plane.

Dynamical model analyses of cycle processes reveal that these models generally include as many poles and zeros at the origin $\left(< 10^{-6}\right)$ of the complex plane as there are components in the refrigerant. The relating states refer to the total masses of species in the cycle which are constant and thus uncontrollable (dosing and venting is disregarded). Consequently, the dynamic order of the model can be reduced by the number of components without affecting model accuracy.

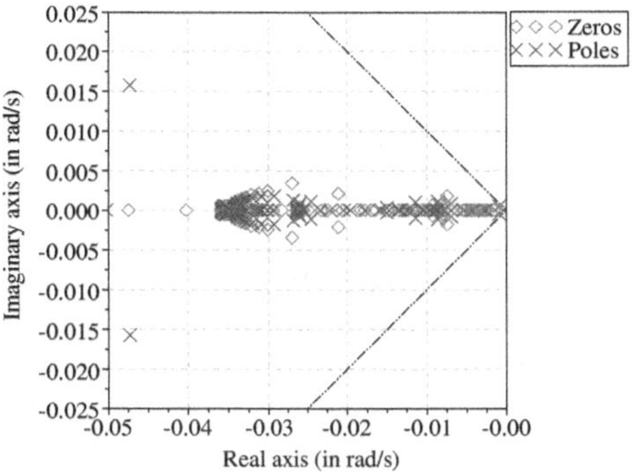

Figure 5.1.: Pole-zero spectrum of the reduced state-space model of the LIMUM® cycle

5.4. LIMUM® cycle

The goal of this section is to compare alternative control strategies of the LIMUM® cycle with respect to their dynamical behavior. Model analysis is done in Section 5.4.1. The control strategies relate to control structures found in Section 4.6 combined with the best I/O pairing which is so far unknown. The best pairing of the CONV and the COMB strategy is identified in Section 5.4.2 and 5.4.3, respectively. In Section 5.4.4 both are compared with each other.

5.4.1. Dynamic model analysis

The same model as in Section 4.6 was considered. The I/O variables of the model together with their span for scaling purposes are given in Table 5.2. By the use of the Pantelides algorithm (Pantelides *et al.*, 1988) it was verified that the DAEs are index free. A linear state-space model was extracted from the OPTISIM® model which consisted of 275 ordinary differential equations. By the use of the Kalman decomposition the minimal realization of the state-space model could be identified. Five controllable/unobservable and 40 uncontrollable/unobservable modes where eliminated. 20 out of 21 poles and zeros close to the origin where not canceled by the Kalman decomposition as they have a considerable relative deviation from each other, *i.e.*, they are distributed in the interval $[-1, 1]\ 10^{-10}\,\text{rad/s}$. Disregarding the modes close to the origin led to a stable reduced model with 610 modes (McMillan degree, Skogestad and Postlethwaite, 2005, Definition 4.3). An extract of the pole-zero spectrum of the reduced model is presented in Figure 5.1. Not shown are poles and zeros with magnitude larger than $\omega_b = 0.05\,\text{rad/s}$. Note that the lower and upper limit of magnitudes of the pole-zero spectrum in the left plane was found to be -1.6 and $-9.6 \cdot 10^{-4}\,\text{rad/s}$, respectively. Two RHPZs at 0.058 and 0.145 rad/s are

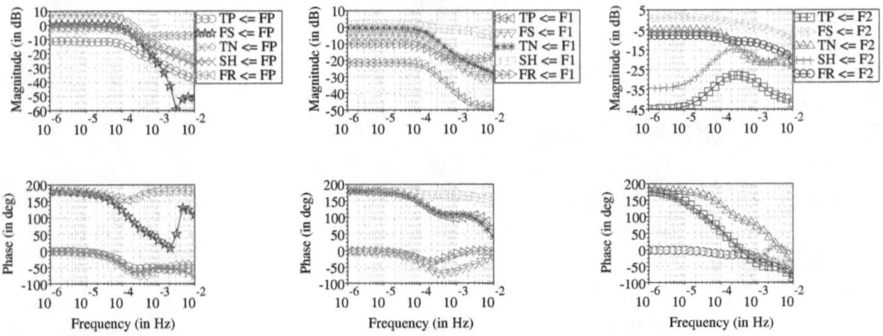

(a) FP (b) F1 (c) F2

Figure 5.2.: Frequency response from MVs to candidate CVs for the LIMUM® cycle

not indicated as they are beyond ω_b and do not impose additional fundamental limitations on control. Figure 5.2 shows the open-loop frequency response from MVs to candidate CVs. It can be concluded that due to fast flow dynamics, the pairs FR $\Leftarrow$ FP, SH $\Leftarrow$ F1 and FS $\Leftarrow$ F1/F2 have good cause and effect over the whole considered frequency band and decouple from other pairs near crossover frequency.

5.4.2. CONV strategy

Due to override control, *i.e.*, either TN or SH is actively controlled, two cases must be considered. Each case has three pairings with negative Niederlinski index. These pairings can be disregarded as they are closed-loop instable. For each case only one pairing of the complete candidate set yields positive RGA elements over the frequency domain. The RGA elements for these pairings are shown in Figure 5.3a and b. Both RGAs are promising for the selected pairings as, after rearrangement, they are close to identity at bandwidth frequency. Further, both pairings have positive Niederlinski index which, together with positive RGA elements, prove that integrity is satisfied (Zhu and Jutan, 1996). Fortunately, the pairings for both cases are compatible with each other, *i.e.*, the CVs involved in the override control are related to the same MV. However, at low frequencies the RGA elements for active TN are relatively large which indicates considerable interactions of the loops at steady-state operation. It can be concluded that the best pairing of the CONV structure is given by

$$\begin{aligned} \mathrm{TP} &\Leftarrow \mathrm{FP} \\ \{\mathrm{TN}, \mathrm{DT}\} &\Leftarrow \mathrm{F1} \\ \mathrm{FS} &\Leftarrow \mathrm{F2} \end{aligned}$$

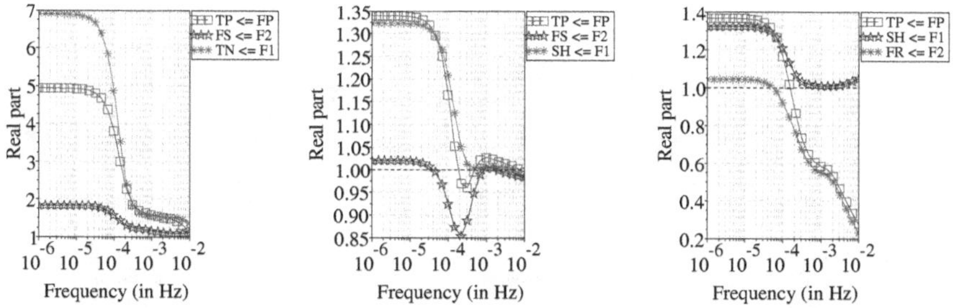

(a) CONV, TN active (b) CONV, SH active (c) COMB

Figure 5.3.: Overall positive RGA elements for control structures of the LIMUM® cycle

and is referred to as the {CONV1, CONV2} strategy, respectively. The selected pairings are further supported from process understanding by their physically closeness and good cause and effect. Moreover, they are practically proven as they have been frequently implemented in the past.

An additional conclusion which can be drawn from the RGA is that the models of both, the CONV1 and CONV2 strategy, do not suffer from uncertainty near bandwidth as the largest RGA entries have magnitudes around one. For the steady-state model it holds almost the same except for some pairings in the CONV1 model which have a larger magnitude but smaller than seven which is considered uncritical.

5.4.3. COMB strategy

As for the CONV structure, there is also only one set of pairings given by

$$\begin{aligned}\mathrm{TP} &\Leftarrow \mathrm{FP}\\ \mathrm{SH} &\Leftarrow \mathrm{F1}\\ \mathrm{FR} &\Leftarrow \mathrm{F2}\end{aligned}$$

which provides positive RGA elements over the considered frequency band. The respective RGA elements of this so called COMB strategy are presented in Figure 5.3c. The COMB strategy does not violate important requirements such as physical closeness and good cause and effect. Unfortunately, two elements of its rearranged RGA diverge from identity at crossover frequency which indicates that fast control may cause interference between two loops. This behavior is due to the above mentioned effect that a high gain from FP towards FR is maintained at bandwidth frequency and thus interferes with the other selected pairs TP$\Leftarrow$ FP and FR$\Leftarrow$ F2.

Besides RGA, other reasons can be identified why other pairings such as SH $\Leftarrow$ F2 and

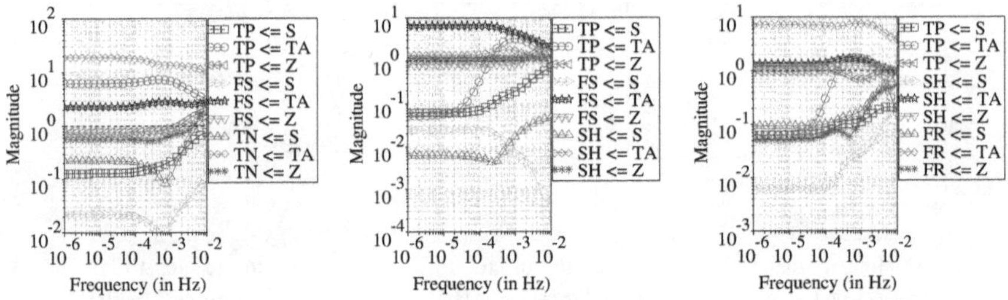

(a) CONV1 (b) CONV2 (c) COMB

Figure 5.4.: Disturbance compensation for the LIMUM® cycle

TP $\Leftarrow$ F2 are forbidden for both, the CONV and the COMB structure. As suggested by the (scaled) frequency response plot in Figure 5.2c, at low frequencies the gain from F2 towards SH and TP is very small indicating saturation of the CJT valve. In contrast to F2 towards TP and SH, all nominated pairings show considerably larger steady-state gains.

Model uncertainty is not an issue as the maximum RGA entries do not significantly exceed the magnitude of one within the investigated domain.

5.4.4. Comparison

Up to frequencies of 10^{-4} Hz, the MSV of the CONV1 model is $2 \cdot 10^{-2}$ whereas the MSVs of the CONV2 and the COMB model are both 10^{-1}. The fact that all MSVs are below one indicates that steady-state MV saturation poses a problem for all strategies (most severely for CONV1). The CNs of the CONV1, the CONV2 and the COMB model are in the order of 10^2, 10^1 and 10^1, respectively, for frequencies smaller 10^{-4} Hz. With one order of magnitude smaller steady-state CNs, the CONV2 and the COMB strategy are also superior over the CONV1 strategy when considering dependence on I/O directions. The CONV1 strategy can be considered almost ill-conditioned.

The steady-state DCNs given by

	S	TA	Z
CONV1	7.7	7.0	1.7
CONV2	2.1	2.1	2.6
FR	3.7	4.4	3.8

indicate that the CONV2 strategy is the least affected in its difficult directions by DVs. The disturbance compensation measure $\boldsymbol{G}^{-1}\boldsymbol{G_d}$ is plotted in Figure 5.4 for all strategies. The requirement $\left(\boldsymbol{G}^{-1}\boldsymbol{G_d}\right)_{ij} < 1$ is in all cases violated when the DV TA is involved.

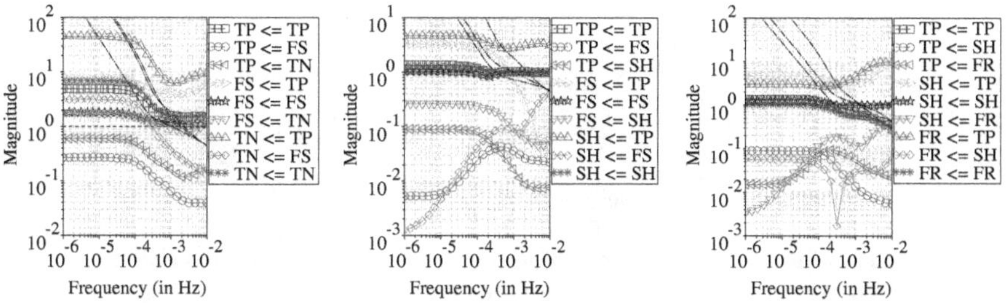

(a) CONV1 (b) CONV2 (c) COMB

Figure 5.5.: PRGAs for the LIMUM® cycle

This suggests that MV constraints pose a problem under influence of ambient temperature. Note that the strongest violation occurs for the CONV1 strategy as, on the one hand, the violation is of strongest magnitude and, on the other hand, all three loops are affected. In contrast, for the CONV2 and COMB strategy only one loop is significantly out of range. The steady-state PRGAs and CLDGs $\left[\, \boldsymbol{\Gamma} \,\middle|\, \tilde{\boldsymbol{G}}_{\boldsymbol{d}} \,\right]$ for the CONV1, CONV2 and the COMB strategy are given by

	TP	TN	FS	S	TA	Z
TP	4.9	−0.6	−0.3	−0.0	1.7	0.2
TN	−44.4	6.9	3.1	0.2	−17.7	−0.5
FS	−7.4	1.8	1.8	0.0	2.7	−0.8

	TP	SH	FS	S	TA	Z
TP	1.3	−0.1	0.0	−0.0	0.0	0.3
SH	−4.7	1.3	−0.0	0.0	1.7	−1.4
FS	3.2	0.3	1.0	0.0	7.5	−0.9

and

	TP	SH	FR	S	TA	Z
TP	1.4	−0.1	0.0	−0.0	0.0	0.3
SH	−4.7	1.3	−0.0	0.0	1.7	−1.4
FR	3.1	0.1	1.0	−0.0	2.9	−0.0

respectively. From the steady-state PRGAs it can be reasoned that for the CONV1 strategy all SVs and CVs have significant interactions, while for the CONV2 and COMB strategy couplings are only dominant for TP setpoint towards all CVs. From steady-state PRGA and GLDG it is expected that the performance in setpoint tracking and disturbance rejection of both the COMB and the CONV2 strategy are superior to that

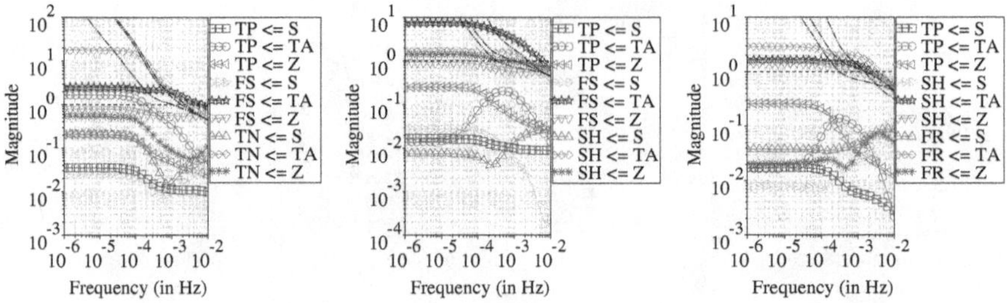

(a) CONV1 (b) CONV2 (c) COMB

Figure 5.6.: CLDGs for the LIMUM® cycle

of the CONV1 strategy. However, more conclusive are the PRGAs and CLDGs near bandwidth frequency which follow from the charts in Figure 5.5 and 5.6, respectively. With few exclusions, all strategies have similar PRGA and CLDG entries near bandwidth frequency. The largest entries of the PRGA near bandwidth are in each case related to the TP setpoint. They do not fall below 1 and consequently put high requirements on the closed-loop bandwidth. Considering the CLDG, it is striking that the DV imposing the highest requirements on the closed loop bandwidth is TA for all cases. At bandwidth, all CLDG entries for all cases are less equal 1. The DV S can be identified as having the least effect on all CVs for all strategies and frequencies.

PID controller tuning was performed based on performance requirements for setpoint tracking and disturbance rejection as proposed in Appendix 5.B. For each case the problem (5.10a) through (5.10c) were solved for every individual loop subsequently. A GM of two and a lower limitation on the integration time of 600 s were imposed. The resulting loop transfer functions $\boldsymbol{L}_{ii}\,\forall i$ are included as guide lines in Figures 5.5 and 5.6. It can be concluded that setpoint tracking and disturbance rejection can be achieved in a time horizon of few hours (corresponding to a closed-loop bandwidth of 10^{-4} Hz and a time constant of a first order lag of 1600 s).

The frequency dependent RDG (not plotted) revealed that the interactions between the loops increase the effect of DVs to CVs more severely for CONV1 than for the CONV2 and the COMB strategy. Most striking is the steady-state increase in the effect from TA towards TP of one order of magnitude for the CONV1 strategy.

All measures used in the evaluation up to now are based on the linear model of the LIMUM® cycle. Consequently, the drawn conclusions must not necessarily hold in general. Therefore, it is necessary to check nonlinear effects with rigorous dynamic simulation in OPTISIM®. This also serves as a final assessment on the quality of the decentralized PI control structure. Four closed-loop dynamic simulation runs were performed. For dy-

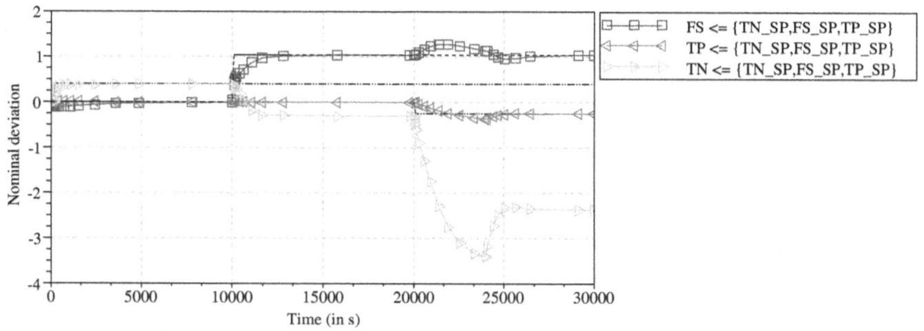

(a) CONV strategy

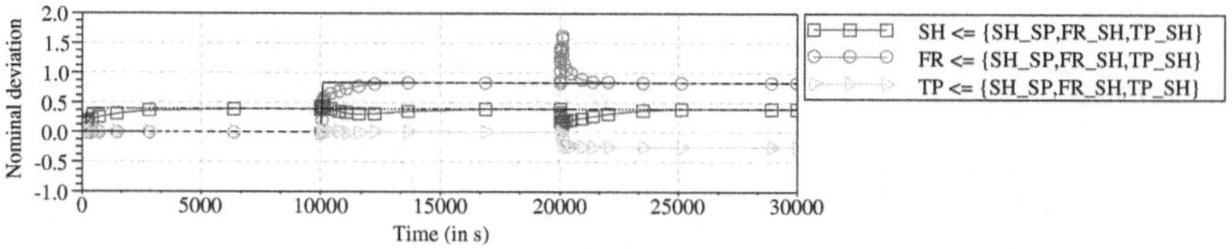

(b) COMB strategy

Figure 5.7.: Scaled simulation results of closed-loop responses due to setpoint steps for the LIMUM® cycle

namic simulations a set of DAEs is solved using a fully implicit backward differentiation formula method. For either strategy the performance of setpoint tracking and disturbance rejection were investigated. The results are presented in Figure 5.7 and 5.8, respectively. For verification of setpoint tracking of the CONV strategy, the TN, FS and TP setpoints were subsequently changed by +2 K, +145 mol/s and −0.5 K, respectively, after intervals of 10^4 s. It was required that SH is always larger than 10 in order to avoid switching from CONV1 to CONV2 and give better comparison between the CONV1 and the COMB strategy. For FS and TP, the setpoint tracking occurs fast and tight. After 5000 s, both setpoints achieve their steady-state target. However, as indicated by the small MSV, the setpoint for TN cannot be held after the second and third setpoint step due to MV saturation of F1 and partly FP. Instead of the setpoint steps of TN and FS, equivalent steps of SH by +2 K and FR by +150, respectively, were performed for the COMB strategy. The imposed setpoints were achieved fast and tightly.

Also after time intervals of 10^4 s, the disturbance variations of TA, Z and S by +4 K, +40 ° and +1 were imposed. As in the setpoint tracking results for the CONV strategy, FS and TP could be kept at their envisaged targets whereas TN significantly dropped out of scope due to MV saturation of F1 and FP. The COMB strategy managed to recover the control targets at steady-state. All transitions were completed in less than 5000 s and verify the estimated time constant of 1600 s. As suggested by the CLDG measure, S

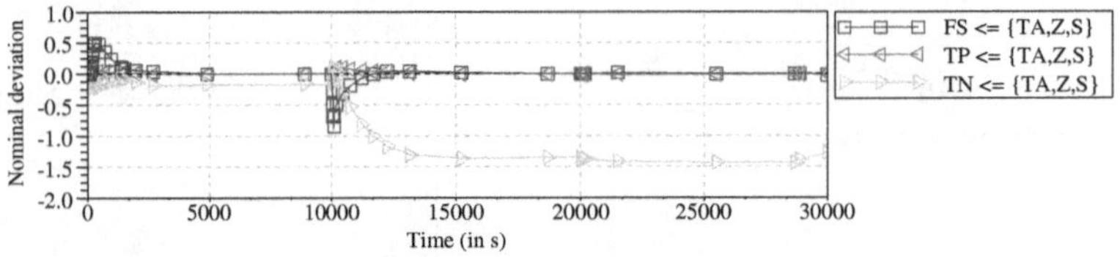

(a) CONV strategy

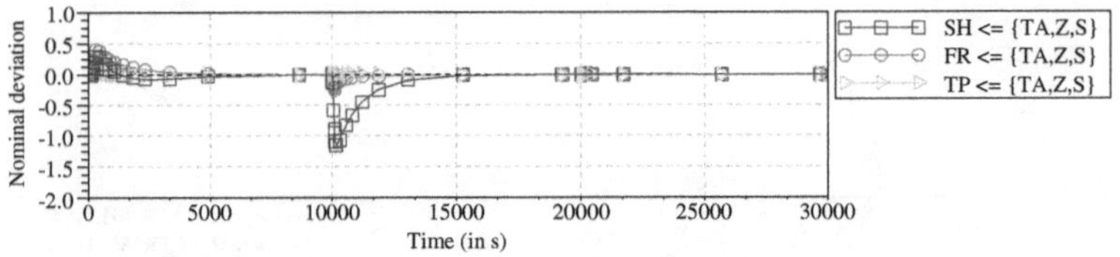

(b) COMB strategy

Figure 5.8.: Scaled simulation results of closed-loop responses due to disturbance affection for the LIMUM® cycle

affects the process only marginally.

5.4.5. Conclusions

From an operability standpoint, the CONV2 and COMB strategy are almost equivalent with the tendency that CONV2 strategy performs slightly better. Both can be considered superior to the CONV1 strategy. What was already indicated from the steady-state considerations is that the targets of controlling TP and TN conflict with each other with respect to the given set of MVs. If HHC separation at this position is necessary, a bypass around the subcooler may be an option for increasing versatility of the process.

5.5. MFC® process

In this section, the dynamic behavior of alternative control strategies of the MFC® process are investigated. The control strategies relate to control structures found in Section 4.7 combined with the best I/O pairing yet to be identified. The best pairing of the CONV and the COMB strategy is identified in Section 5.5.2 and 5.5.3. In Section 5.5.4 both are compared with each other.

5.5.1. Dynamic model analysis

The dynamic model of the MFC® process discussed here agrees with the one introduced in Section 4.7. Two control structures are compared, the CONV and the COMB structure.

Type	Var.	Description	Nom. point	Span
MVs	FP	LNG flow rate (in mol/s)	27411	2700
	F1	PC flow rate (in mol/s)	36172	3600
	F2	LC flow rate (in mol/s)	26464	2600
	F3	SC flow rate (in mol/s)	17572	1700
	C1	PC compression ratio	3.6	0.7
	C2	LC compression ratio	6.2	1.2
	C3	SC compression ratio	13.4	2.7
DVs	N	Compressor speed (in RPM)	3000	90
	TA	Ambient temperature (in K)	298.15	10.0
Cand. CVs	TP	LNG temperature (in K)	109.3	2.0
	TN	Precooled natural gas temp. (in K)	256.5	5.0
	T3	SC temp. after liquefier (in K)	202.9	5.0
	F3	$\rightarrow$ MVs		
	P1	PC discharge pressure (in bar)	23.2	2.3
	P2	LC discharge pressure (in bar)	25.5	2.6
	P3	SC discharge pressure (in bar)	46.2	4.6
	S1	PC superheating (in K)	10.0	5.0
	S2	LC superheating (in K)	10.0	5.0
	S3	SC superheating (in K)	10.0	5.0
	FR	PC & LC flow rate relation (mol/s)	9707.9	1000
	PL	LC pressure relation (in bar)	-1.72	2.6
	PS	SC pressure relation (in bar)	6.96	4.6

Table 5.3.: I/O variables of the state-space model of the MFC® cycle

MVs, DVs and CVs of the process at nominal operating point including their span for scaling purposes are summarized in Table 5.3.

The Pantelides algorithm (Pantelides *et al.*, 1988) applied to the OPTISIM® model revealed that the model is index-free. The OPTISIM® model was exported into a linear state-space model with 722 ordinary differential equations. Three uncontrollable/observable and 95 uncontrollable/unobservable modes were identified by the application of the Kalman decomposition. It is noteworthy that each of the three cycles carries seven species yielding 21 uncontrollable/observable modes referring to the invariant total charge of each species. Accordingly, 18 of 21 pole-zero pairs near the origin were not identified due to their relative offset of several orders of magnitude (their absolute magnitudes were distributed within the interval $\left[10^{-17}, 10^{-11}\right]$ rad/s). This is due to the fact that the solution of the OPTISIM® model has a certain numerical accuracy which is transferred to the linear state-space model. Disregarding the pole/zero pairs at the origin led to a McMillan degree of 704. An extract of the pole/zero spectrum in the practically relevant interval $[0, \omega_b]$ is show in Figure 5.9. The complete pole/zero spectrum in the left hand plane is distributed within the interval $\left[-10.03, -3.45 \cdot 10^{-3}\right]$ rad/s. Two RHPZs occur

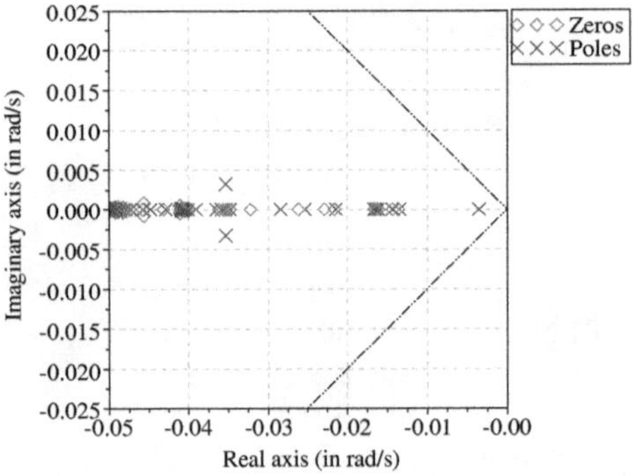

Figure 5.9.: Pole-zero spectrum of the reduced state-space model of the MFC® process

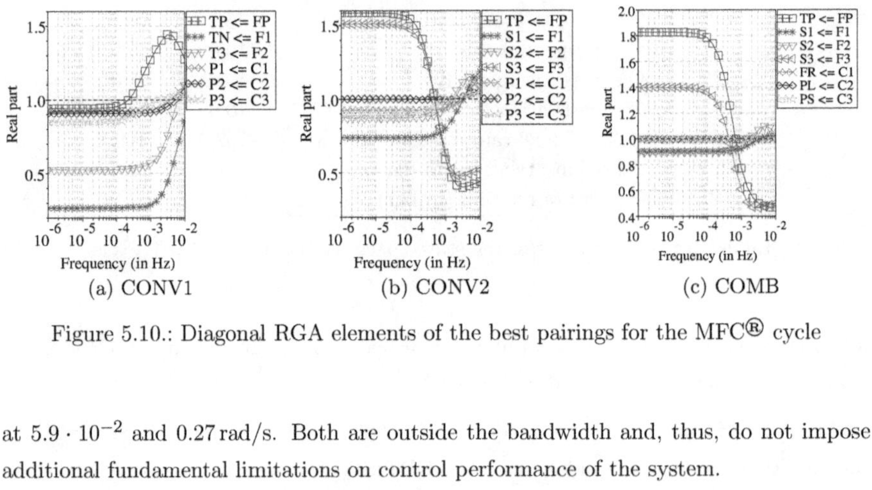

(a) CONV1 (b) CONV2 (c) COMB

Figure 5.10.: Diagonal RGA elements of the best pairings for the MFC® cycle

at $5.9 \cdot 10^{-2}$ and 0.27 rad/s. Both are outside the bandwidth and, thus, do not impose additional fundamental limitations on control performance of the system.

5.5.2. CONV

The CONV structure possesses three override control loops to ensure that compressor inlet is always above a minimum superheating bound. Thus, there are $\sum_{i=1}^{3} C_i^3 = 8$ structural combinations which need to be verified. For the sake of simplicity, only boundary cases are considered here, *i.e.*, the CONV1 case in which all loops are in regular operation and the CONV2 case in which all loops are overridden such that S1, S2 and S3 are active. From the frequency dependent RGA of the CONV1 and the CONV2 structure the practically proven pairings

$$\begin{aligned}
\mathrm{TP} &\Leftarrow \mathrm{FP} \\
\{\mathrm{TN}, \mathrm{S1}\} &\Leftarrow \mathrm{F1} \\
\{\mathrm{T3}, \mathrm{S2}\} &\Leftarrow \mathrm{F2} \\
\{\mathrm{F3}, \mathrm{S3}\} &\Leftarrow \mathrm{F3} \\
\mathrm{P1} &\Leftarrow \mathrm{C1} \\
\mathrm{P2} &\Leftarrow \mathrm{C2} \\
\mathrm{P3} &\Leftarrow \mathrm{C3}
\end{aligned}$$

can be satisfactorily confirmed. The diagonal entries of the rearranged RGA are indicated in Figure 5.10a and b. For the CONV1 structure, there is only one reasonable alternative, given by T3 $\Leftarrow$ F1, TN $\Leftarrow$ F2 which provides better steady-state RGA entries. However, the entries drop sharply towards zero at bandwidth and are thus not favorable.

The open-loop frequency responses for the selected pairings are presented in Figure 5.11a and b. Those responses with frequency invariant gains or no I/O effect are ignored. The pair TP $\Leftarrow$ FP is the only which shows a significant lag, *i.e.*, a gain reduction near bandwidth. This is reasonable as both, compression ratio and refrigerant flow rates, have fast effects on suction pressure and the suction pressure in turn has a fast effect on the shell-side refrigerant temperatures in all SWHE bundles simultaneously. The sensible heat of the SWHE metal has a minor effect on the shell-side temperatures. In contrast, the effect of FP towards TP is slow as only heat transfer and heat capacity rate of the natural gas stream are affected which take effect initially at the inlet of the first SWHE and propagate through all subsequent SWHE bundles with a velocity determined by mass and heat capacity of the metal. The I/O gains corresponding to the MVs F1 and F2 are considerably smaller for CONV1 than for CONV2. Also interesting to remark is the fact that the I/O gains of equivalent pairs, *e.g.*, P1 $\Leftarrow$ C1 and P2 $\Leftarrow$ C2, show similar frequency responses in terms of magnitude and phase.

The CONV1 and the CONV2 structure including the selected pairings are referred to as CONV1 and CONV2 strategy, respectively, in the following. For both models, CONV1 and CONV2, model uncertainty is small as none of the RGA entries' magnitudes grow larger than two.

5.5.3. COMB

From the frequency dependent RGA only one reasonable pairing combination can be concluded. It is given by

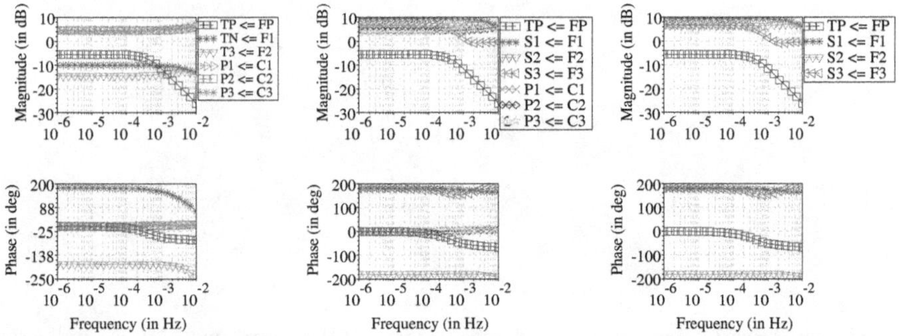

(a) CONV1 (effects of F3 disregarded)

(b) CONV2

(c) COMB (effects of C1, C2, C3 disregarded)

Figure 5.11.: Frequency response from MVs to candidate CVs for the MFC® process

$$
\begin{aligned}
TP &\Leftarrow FP \\
S1 &\Leftarrow F1 \\
S2 &\Leftarrow F2 \\
S3 &\Leftarrow F3 \\
FR &\Leftarrow C1 \\
PL &\Leftarrow C2 \\
PS &\Leftarrow C3
\end{aligned}
$$

and is referred to as COMB strategy. The open-loop frequency responses of the diagonal elements of the COMB strategy which are frequency dependent are shown in Figure 5.11c. They are very similar to those of the CONV2 strategy.

It is noteworthy that model uncertainty is not an issue as all RGA elements are smaller than two over the considered frequency band.

5.5.4. Comparison

The steady-state MSVs for all strategies are in the order of 10^{-1}. While the MSV declines towards 10^{-2} at bandwidth for the CONV1 strategy, it is almost frequency independent for the CONV2 and the COMB strategy. Consequently, steady-state MV saturation states a problem for all strategies. The CN of the CONV1 strategy is 10^1 at steady-state and rises to 10^2 near bandwidth. Accordingly, strong dependency on plant directions is only expected at small time scales. The CONV2 and COMB strategy have frequency

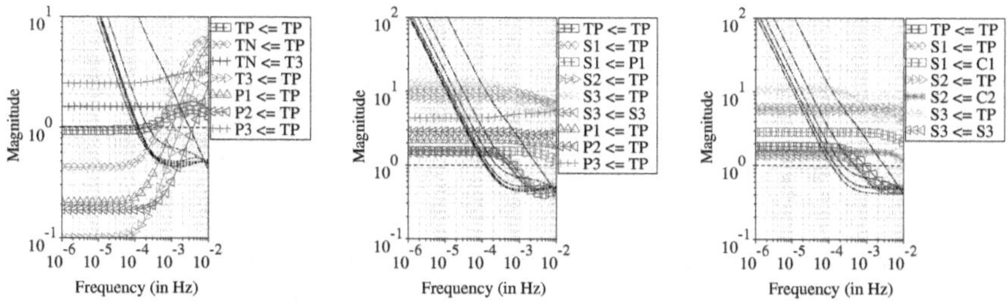

(a) CONV1 (b) CONV2 (c) COMB

Figure 5.12.: PRGAs for the MFC® process

independent CNs of $5 \cdot 10^1$ and $3 \cdot 10^1$, respectively, and are thus not considered ill-conditioned.

The steady-state DCN for the three strategies are given by

	N	TA
CONV1	2.3	4.1
CONV2	4.3	2.7
CRC	4.9	1.7

where the values for the CONV1 strategy tend to higher values towards bandwidth and vice versa for the CONV2 and the COMB strategy. The DCN figures suggest that none of the strategies are severely affected in their difficult directions by DVs. For all strategies, the disturbance compensation measure $\boldsymbol{G}^{-1}\,\boldsymbol{G_d}$ has only one element which is significantly larger than 1 within the frequency band. The pair TN $\Leftarrow$ TA (CONV1 strategy) has a value of 4.5 and the pair S1 $\Leftarrow$ TA (CONV2 and COMB strategy) has a value of 2.6 and 1.5.

Those PRGA and CLDG elements which are significantly above one within the frequency band are indicated in the charts of Figures 5.12 and 5.13, respectively. For all strategies, the majority of PRGA/CLDG elements have magnitudes below one and it can thus be expected that setpoint tracking and disturbance rejection performance is relatively good. Steady-state loop interactions are more dominant for the CONV2 and the COMB strategy than for the CONV1 strategy as the magnitudes of the largest PRGA elements vary by one order of magnitude. However, near bandwidth this difference vanishes as interactions for CONV1 increase. All in all, one can say that the strong interactions are dominated by the TP loop for all strategies. For the CONV1 strategy, the magnitude of only one CLDG entry, namely that of the pair TN $\Leftarrow$ TA, is marginally above one. In contrast for the CONV2 and the COMB strategy, several entries are more significantly

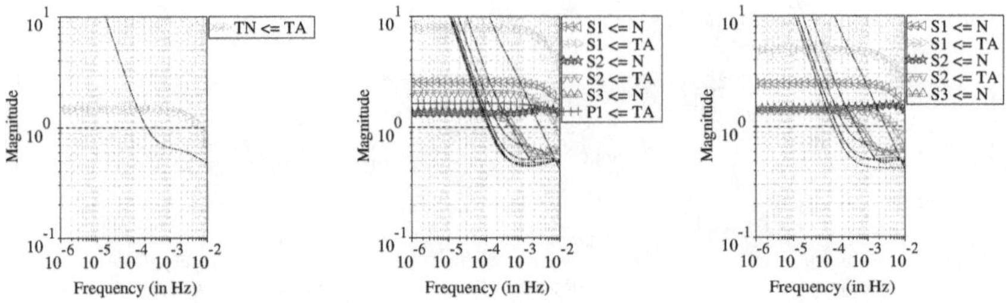

(a) CONV1 (b) CONV2 (c) COMB

Figure 5.13.: CLDGs for the MFC® process

larger than one. It is interesting to note that for both, the CONV2 and the COMB strategy, the S1 and the S2 loops are majorly affected by the disturbances.

Also included in the PRGA and CLDG plots are loop transfer functions $\boldsymbol{L}_{ii} \forall i$. The respective PID controller parameters were calculated based on performance requirements for setpoint tracking and disturbance rejection as proposed in Appendix 5.B. For each case the problem (5.10a) through (5.10c) was solved for every individual loop subsequently. A GM of two and a lower limitation on the integration time of 600 s were imposed. For all strategies, the largest loop transfer function relates to the CV TP. For the CONV1 strategy, it can be concluded that a closed-loop bandwidth around 10^{-4} Hz can be achieved (corresponding to a time constant of first order lag of 1600 s) as at smaller frequencies all loop transfer functions have larger magnitudes than their related PRGA and CLDG entries. For the CONV2 and the COMB strategy a similar conclusion can be drawn although some of their PRGA and CLDG entries are slightly larger. The three highest PRGA elements refer to loop TP which has the highest loop transfer function and can thus be compensated at a bandwidth of 10^{-4} Hz. The remaining PRGA elements are crossed by the remaining transfer loops near the same frequency. However, the effect of both disturbances on S1 can only be compensated at somewhat smaller bandwidths than for the CONV1 strategy, *i.e.*, somewhere in between 10^{-5} and 10^{-4} Hz.

The RDG (not plotted) was inspected for all strategies in order to check how the loop interactions affect the disturbance gains. For the CONV1 strategy, it was figured out that the loop interactions reduce the disturbance effect for almost every DV/CV pair. Those who are not reduced are only increased marginally by less than factor 1.5. For both, the CONV2 and the COMB strategy, most of the RDG elements in the column for N are between one and two. The largest element is the one of the pair S3 $\Leftarrow$ N and has a magnitude of 10. Fortunately, the absolute disturbance gain of this structure is comparably small as may be seen in Figures 5.13b and c.

In order to consider nonlinear effects and for the verification of the decentralized PI controller design, rigorous dynamic simulation of the closed-loop plants were performed in OPTISIM®. The two DVs and, for the sake of reducing the effort, only two setpoint variables were altered for each strategy. For the CONV strategy, the LNG setpoint temperature (TP_SP) was increased by 1 K, the setpoint of the natural gas temperature after the precooler (TN_SP) was increased by 2 K, the ambient temperature (TA) was increased by 5 K and the compressor speed N was increased by 60 RPM. The simulation result is presented in Figure 5.14a. F3 is a feed-forward variable and thus not indicated. As the safety margins for superheating have not been violated the control strategy was not switched and only the CONV1 strategy was active. As suggested by the controllability measures, setpoint tracking and disturbance rejection shows very good performance. In less than 5000 s after the setpoint or disturbance affection (three times the proposed time constant), the control goals can be considered as recovered.

For the COMB strategy, a similar experiment was performed with the difference that instead of the setpoint step in the natural gas temperature, the setpoint of the superheating temperature of the PC mixed refrigerant was increased by 2 K. The simulation result is presented in Figure 5.14b. As suggested by the PRGA plot in Figure 5.12c, there are significant loop interactions between TP and S1 through S3. This can be verified by observation of step response caused by the affection of TP_SP. While the setpoint of TP is achieved fast and tightly, S1 through S3 are considerably deviated and need three times longer for returning to their nominal values as in the CONV1 strategy.

5.5.5. Conclusions

It was shown that from a controllability standpoint, the CONV1 strategy shows acceptable dynamical behavior and performs slightly better than the CONV2 and the COMB strategy. Comparing the COMB with the CONV strategy (CONV2 overrides CONV1), the difference in the dynamic behavior is almost negligible. From the author's point of view, the loss in dynamical performance of the COMB strategy is more than compensated by the economic advantage outlined in section 4.7.

Although plant scale between the LIMUM® and the MFC® case study varied dramatically (factor 40), the plant dynamics are in the same order of magnitude. This indicates that competing effects on plant dynamics compensate each other. *E.g.*, equipment metal mass grows proportionally with the heat capacity rate of the natural gas and mixed refrigerant streams.

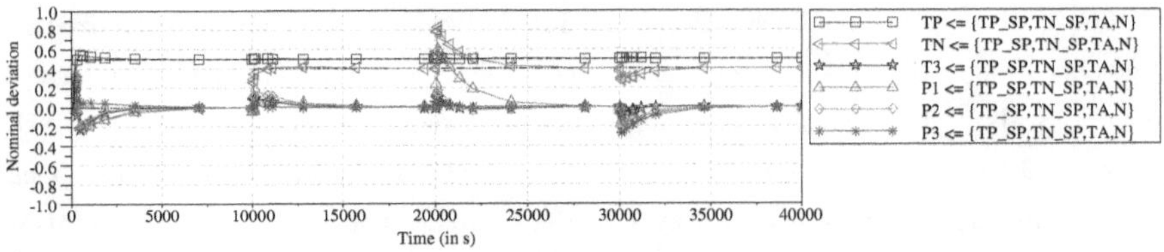

(a) CONV strategy

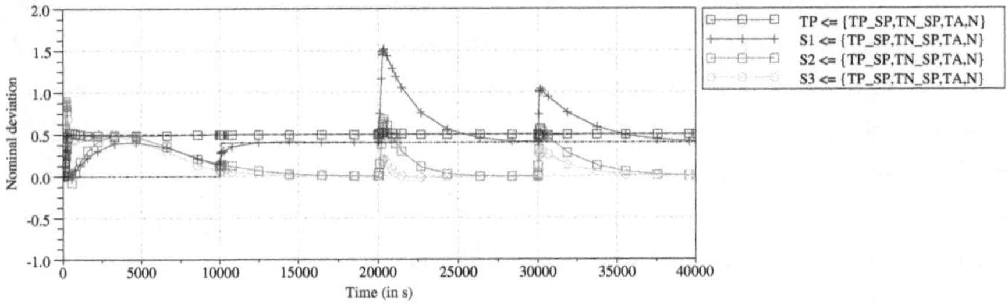

(b) COMB strategy

Figure 5.14.: Scaled simulation results of closed-loop responses due to setpoint tracking and disturbance affection for the MFC® process

5.6. Conclusions

It was proven by means of dynamical measures that newly proposed control structures for the LIMUM® cycle (Section 4.6) and the MFC® process (Section 4.7) can serve as proper replacements of conventional control structures. For each structure, the best pairing in terms of interactions and integrity have been identified yielding the corresponding control strategy. It was figured out that the COMB strategy for the LIMUM® cycle can be seen as a better replacement due to better setpoint tracking and disturbance rejection performance. Due to override control, the conventional strategy of the MFC® process consists of eight structurally different configurations and is therefore difficult to assess. To keep the effort in check, only two structural alternatives representing the boundary cases were investigated. It was figured out that the newly proposed strategy performs similarly but not quite as good as the conventional strategy.

5.A. Operability measures

Before operability measures can be introduced, the concept of control needs to be clarified. Suppose the plant is represented by the linear model

$$\boldsymbol{c} = \boldsymbol{G}(s)\ \boldsymbol{u} + \boldsymbol{G}_d(s)\ \boldsymbol{d} + \boldsymbol{n}^c.$$

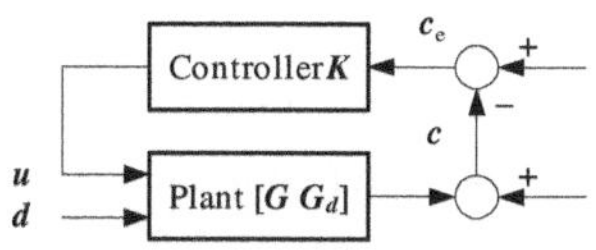

Figure 5.15.: Controlled plant

The I/O matrices of the plant in the frequency domain can be obtained from a linear state-space model via the transformation

$$\begin{bmatrix} \boldsymbol{G}(s) & \boldsymbol{G_d}(s) \end{bmatrix} = \boldsymbol{C}\,(s\boldsymbol{I}-\boldsymbol{A})^{-1}\,\boldsymbol{B}+\boldsymbol{D}.$$

The plant is controlled by feedback controller $\boldsymbol{K}(s)$ as illustrated in Figure 5.15. By simple algebra the closed-loop response of the control error

$$\boldsymbol{c}_{\mathrm{e}} = \boldsymbol{S}(s)\,\boldsymbol{c}_{\mathrm{s}} + \boldsymbol{S}(s)\,\boldsymbol{G_d}\,\boldsymbol{d} + \boldsymbol{T}(s)\,\boldsymbol{n}^{\mathrm{c}} \tag{5.1}$$

can be derived. Here, the sensitivity and complementary sensitivity function are given by $\boldsymbol{S}(s) = (\boldsymbol{I}+\boldsymbol{L}(s))^{-1}$ and $\boldsymbol{T}(s) = \boldsymbol{I} - \boldsymbol{S}(s) = (\boldsymbol{I}+\boldsymbol{L}(s))^{-1}\,\boldsymbol{L}(s)$, respectively, where $\boldsymbol{L}(s) = \boldsymbol{G}(s)\,\boldsymbol{K}(s)$ is referred to as the loop transfer function. For the sake of simple notation, the indication of the dependency on s is dropped in the sequel.

The ideal case would be to have $\boldsymbol{S} = \boldsymbol{0}$, *i.e.*, vanishing setpoint deviation and perfect disturbance rejection, and $\boldsymbol{T} = \boldsymbol{0}$, *i.e.*, noise elimination. As $\boldsymbol{T}+\boldsymbol{S} = \boldsymbol{I}$ always holds, both goals are contrary and for controller design a compromise between both must be made. Fortunately, noise and control usually act on different non-overlapping time scales for chemical plant operation. *I.e.*, at short time scales (high frequencies) where noise is present, a chemical plant usually reacts inertly and control is needless. Thus, it is desirable to have $\lim_{\omega\to\infty}\boldsymbol{T} = \boldsymbol{0}$ which is always provided as every real system is strictly proper (Skogestad and Postlethwaite, 2005, pp. 81). At long time scales (low frequencies), it is desired that control is active, corresponding to $\lim_{\omega\to 0}\boldsymbol{S} = \boldsymbol{0}$. According to Skogestad and Postlethwaite (2005, pp. 38, 81), control is considered effective as long as the relative error $\|\boldsymbol{c}_{\mathrm{e}}\|_2 \,/\, \|\boldsymbol{c}_{\mathrm{s}}\|_2 \leq \bar{\sigma}(\boldsymbol{S})$ is reasonably small, precisely if $\bar{\sigma}(\boldsymbol{S}) \leq 2^{-1/2}$. The smallest frequency at which $\bar{\sigma}(\boldsymbol{S})$ crosses $2^{-1/2}$ from below is referred to as the bandwidth ω_{b}. At frequencies below the bandwidth, the plant can be controlled tightly as changes in both setpoints and disturbances can be compensated for. In general, the larger the bandwidth of a system the faster the time response but the more vulnerable for noise it is. There are some other characteristic frequencies namely ω_{c}, the crossover frequency, and $\omega_{\mathrm{b}T}$, the bandwidth in terms of T. They are both defined for single-input-single-output (SISO) systems (Skogestad and Postlethwaite, 2005, pp. 38-40) but are sometimes used interchangeably with ω_b as they are located close to each other.

All mentioned characteristic frequencies are properties of a controlled plant and thus depend on controller design. However, properties of the open loop system impose restrictions on the achievable bandwidth and it is thus possible to predict closed-loop behavior by studying the open loop plant. For this purpose, various measures have been proposed. An important selection is briefly recalled in the sequel. It is important to stress that with few exceptions, the most measures are particularly meaningful when they are evaluated in the frequency domain near the bandwidth, *i.e.*, the upper frequency bound at which control is supposed to be active. Therefore, it is particularly valuable to have a good model of time delays and dynamics near bandwidth, especially RHPZ (Wolff, 1994, p. 9).

Remark 5.1. I/O scaling is crucial for proper interpretation of the most measures. Scaling of the I/O variables is performed via multiplication of the augmented plant $\begin{bmatrix} \boldsymbol{G} & \boldsymbol{G_d} \end{bmatrix}$ with diagonal scaling matrices, *i.e.*, diag $\left(1/\Delta_1^{\boldsymbol{y}}, \ldots\right)$ from the left and diag $\left(\Delta_1^{\boldsymbol{u}}, \ldots, \Delta_1^{\boldsymbol{d}}, \ldots\right)$ from the right, where $\Delta_i^{\boldsymbol{v}}$ represents the span of the respective variable $\boldsymbol{v}_i$, such that the scaled variable $\boldsymbol{v}_i/\Delta_i^{\boldsymbol{v}}$ is within the interval $[-1, 1]$. In this work, it is assumed that the allowed variations of $\boldsymbol{u}$, $\boldsymbol{d}$ and $\boldsymbol{y}$ do not depend on frequency and therefore their spans can be considered frequency independent.

5.A.1. RHPZ

A zero of a multivariate system refers to a frequency z where the transfer function $\boldsymbol{G}(s)$ or the Rosenbrock matrix of a state-space system $\boldsymbol{P}(s)$ loses rank. More practically spoken, if a system is stimulated by an MV transient which has frequency z, then the CVs' signals show linear dependency which means that the CVs cannot be independently controlled from each other anymore. This effect is due to internal competing effects which may even cause CV signal distinction as in the SISO case where the rank drops from 1 to 0. It is important to note that the problem

$$\boldsymbol{P}(z) \begin{bmatrix} \boldsymbol{x}_z \\ \boldsymbol{u}_z \end{bmatrix} = \boldsymbol{0}$$

can be transferred into a generalized eigenvalue problem which may include solutions at infinity. However, only finite solutions are considered zeros of the system. It must further be remarked that zeros of multivariate systems are sometimes called transmission/multivariable zeros in order to distinguish them from zeros of a transfer function of an individual I/O pair.

Right hand plane zeros (RHPZs) refer to zeros in the right half of the complex plane. Multivariable RHPZs of a plant fundamentally limit its achievable bandwidth regardless of the applied controller. One reason for this effect is that perfect control is not possible as the controller would need an instable pole for RHPZ compensation which is practically infeasible. RHPZ in the transfer functions of the diagonal pairs of the rearranged plant are obstructive if decentralized controllers are used, but generally do not pose a problem for multivariate controllers. The closer an RHPZ to the origin the more severe is its effect. The closest RHPZ $\tilde{z}$ dominates the upper limit on the achievable bandwidth, *i.e.*, $\omega_\mathrm{b} < \tilde{z}$. Multivariable RHPZ are always related to a certain I/O direction and are thus less severe as RHPZ of SISO systems, especially if they lie in CV directions where performance requirements are uncritical (Morari *et al.*).

5.A.2. CN and DCN

The condition number (CN) of a scaled plant $\boldsymbol{G}$ is defined as

$$\mathrm{CN} = \bar{\sigma}\left(\boldsymbol{G}\right)/\underline{\sigma}\left(\boldsymbol{G}\right),$$

where $\bar{\sigma}$ and $\underline{\sigma}$ denotes the largest and smallest singular value, respectively. The CN indicates the sensitivity from I/O directions to I/O magnitude. For instance, if the CN is far way off from its lower bound 1, then an MV variation in one particular direction has a tiny effect on the CVs (in terms of 2-norm) while in another particular direction the CV variance may be huge. A plant with a high CN is said to be ill-conditioned and requires widely different MV magnitudes depending on the direction of the desired CVs.

Skogestad and Morari (1987) introduced the disturbance condition number (DCN) for DV i,

$$\mathrm{DCN}_i = \frac{\left\|\boldsymbol{X}^{-1}\left(\boldsymbol{G_d}\right)_i\right\|_2}{\left\|\left(\boldsymbol{G_d}\right)_i\right\|_2}\,\bar{\sigma}\left(\boldsymbol{X}\right),$$

where $\boldsymbol{X}$ may be the open-loop gain $\boldsymbol{G}$ or the loop transfer function $\boldsymbol{L}$ and $\left(\boldsymbol{G_d}\right)_i$ indicates the ith row of $\boldsymbol{G_d}$. The disturbance condition number DCN_i reveals how severe the DV i affects the plant in its difficult directions. In other words, it represents the ratio of MV magnitudes for disturbance rejection between the cases when the ith unit disturbance acts in the best and in the worst direction (corresponding to the CN). The lower bound of the DCN_i is 1 and relates to an easy-to-reject DV. As the CN, the DCN depends on scaling.

5.A.3. Niederlinski index

Niederlinski (1971) proposed a steady-state measure for stability of decentralized controlled plants, the Niederlinski index

$$\mathrm{NI} = \det \boldsymbol{G}_0 / \prod_i \left(\boldsymbol{G}_0\right)_{ii}. \tag{5.2}$$

The open loop steady-state plant model $\boldsymbol{G}_0$ in (5.2) needs to be rearranged such that the selected pairings relate to the diagonal entries. A positive NI states a necessary but not a sufficient condition for stability. The NI is therefore useful for excluding candidate pairings. The NI is independent of controller parameters. The only necessary assumptions for derivation of (5.2) is that (i) the decentralized controllers must include integral behavior (mostly the case in process industry), (ii) the open loop system is stable and (iii) each loop is stable subject to all other loops open.

5.A.4. RGA

Bristol (1966) introduced an important measure of interaction in decentralized controlled multivariate systems, the relative gain array (RGA) with the compact expression

$$\boldsymbol{\Lambda} = \boldsymbol{G} \otimes \boldsymbol{G}^{-T}. \tag{5.3}$$

Here, $\otimes$ denotes elementwise multiplication (Hadamard/Schur product). The RGA was originally defined for steady-state, but was extended to the frequency domain by McAvoy (Sep 1983) who also suggests referring to (5.3) as the Bristol number. It is important to stress that the right hand side of the product is the regular (not the conjugate) transpose. From (5.3) it is evident that the RGA is only defined for square plants, but (5.3) can be generalized for non-square ones (Skogestad and Postlethwaite, 2005, p. 528). Critics of the RGA argue that the inverse of the plant may be nonproper and a physical interpretation in terms of perfect control is not strictly meaningful except at steady-state. The RGA has some interesting algebraic properties which are omitted here for the sake of brevity. The reader is referred to the work of Skogestad and Hovd (1990/05/23-25) or standard multivariate control textbooks, *e.g.*, Skogestad and Postlethwaite (2005, p. 527). More important is the interpretation of the RGA and the pairing rule which can be derived thereof.

Definition 5.2. The $\{i, j\}$ entry of the RGA is measure how the open-loop gain of the related I/O pairing (*i.e.*, $\boldsymbol{u}_k = 0 \, \forall k \neq j$) is affected by closing other loops under the assumption of perfect control[2] (*i.e.*, $\boldsymbol{y}_k = 0 \, \forall k \neq i$). This can be denoted by

$$\boldsymbol{\Lambda}_{ij} = \frac{\partial \boldsymbol{y}_i / \partial \boldsymbol{u}_j|_{\boldsymbol{u}_k = 0 \forall k \neq j}}{\partial \boldsymbol{y}_i / \partial \boldsymbol{u}_j|_{\boldsymbol{y}_k = 0 \forall k \neq i}} = \frac{\boldsymbol{G}_{ij}}{1/\left(\boldsymbol{G}^{-1}\right)_{ji}}.$$

It is preferred that the open loop gains of selected pairings of a decentralized control structure are as least as possible affected by closing other loops. Such non-interacting

[2]Note that the achievement of perfect control may only be sufficient for frequencies smaller than the bandwidth.

pairings refer to RGA entries close to one. The more off from one an RGA entry[3], the more undesirable is the selection of the relating pair. If the off-diagonal gains of a (rearranged) plant are on both sides of the diagonal non-zero, then the plant has a non-identity RGA and is called interactive. It is known that interactive plants with large RGA elements are ill-conditioned and close to singular irrespective of I/O scaling. According to Skogestad and Hovd (1990/05/23-25), these plants are fundamentally different to control and it is not recommended to use inverse-based controllers (decouplers) for such plants as extreme sensitivity to MV uncertainty is expected. The most unwanted situation occurs when the gains of particular pairings change sign between open and closed-loop configuration. This violates integrity as instable operation of the plant may be caused by controller saturation or deactivation (due to failure or manual mode operation).

Rule 5.3. *Prefer pairings such that the rearranged system, with the selected pairings along the diagonal, has an RGA close to identity at frequencies around the closed-loop bandwidth. For a stable plant avoid pairings that corresponds to negative steady-state RGA elements (Skogestad and Postlethwaite, 2005, p. 450).*

Apart from pairing selection and interaction, the RGA provides further valuable information. As pointed out by Skogestad and Hovd (1990/05/23-25), small values of the RGA entries indicate insensitivity to modeling error.

5.A.5. PRGA, CLDG and RDG

While the RGA is very good for selecting promising pairings, it lacks for appropriately measuring interactions. This is due to the fact that one-way interacting plants, *i.e.*, plants with upper/lower triangular I/O gain, possess an RGA equal to identity. The performance relative gain array (PRGA) by Hovd and Skogestad (1992),

$$\boldsymbol{\Gamma} = \tilde{\boldsymbol{G}}\boldsymbol{G}^{-1}, \tag{5.4}$$

is a better measure particularly for one-way interactions. For evaluation of (5.4), $\boldsymbol{G}$ must be rearranged such that the selected pairings relate to the diagonal elements. The diagonal matrix $\tilde{\boldsymbol{G}}$ possesses the same diagonal entries as $\boldsymbol{G}$. Observation of (5.4) reveals that the PRGA can be seen as a normalized plant inverse. The diagonal entries of the PRGA agree with that of the RGA and the off-diagonals depend on CV scaling. The expression (5.4) can be motivated by considering the factorization

$$\boldsymbol{S} = \left(\boldsymbol{I} + \tilde{\boldsymbol{S}}\left(\boldsymbol{\Gamma} - \boldsymbol{I}\right)\right)^{-1} \tilde{\boldsymbol{S}}\boldsymbol{\Gamma}. \tag{5.5}$$

[3]For nondiscriminatory judgment in each direction, it is recommended to measure the distance on a log scale.

Here, the diagonal matrix $\tilde{\boldsymbol{S}} = \left(\boldsymbol{I} + \tilde{\boldsymbol{G}}\,\boldsymbol{K}\right)^{-1}$ has the important property that $\tilde{\boldsymbol{S}} \approx \boldsymbol{0}$ if control is effective, which holds for frequencies below bandwidth. According to (5.5), it holds $\boldsymbol{S} \approx \tilde{\boldsymbol{S}}\,\boldsymbol{\Gamma}$ in this frequency band and (5.1) can be written as

$$\boldsymbol{c}_{\mathrm{e}} = \tilde{\boldsymbol{S}}\,\boldsymbol{\Gamma}\,\boldsymbol{c}_{\mathrm{s}} + \tilde{\boldsymbol{S}}\,\boldsymbol{\Gamma}\,\boldsymbol{G}_d\,\boldsymbol{d} + \left(\boldsymbol{I} - \tilde{\boldsymbol{S}}\,\boldsymbol{\Gamma}\right)\,\boldsymbol{n}^{\mathrm{c}}. \tag{5.6}$$

Consequently, larger than one off-diagonal PRGA entries mean that the interactions interfere each other and slow down the overall response in comparison to individual loop performance. Vice versa, small off-diagonal PRGA entries in comparison to one indicate that the related loops support each other.

Equation (5.6) gives a clue to another performance measure, the closed-loop disturbance gain (CLDG)

$$\tilde{\boldsymbol{G}}_d = \boldsymbol{\Gamma}\,\boldsymbol{G}_d, \tag{5.7}$$

which relies on DV and CV scaling. The CLDG is useful for determining the necessary bandwidth for disturbance rejection. *I.e.*, in order to reject the ith DV within the desired bandwidth, the elements in the ith column of $\tilde{\boldsymbol{S}}\,\tilde{\boldsymbol{G}}_d$ must be smaller than one for frequency up to bandwidth. Accordingly, pairings which relate to smaller CLDG entries are preferred.

A convenient measure for identifying the DV/CV pairs whose gains are amplified or debilitated by interactions is the relative disturbance gain (RDG) by Stanley *et al.* (1985),

$$\tilde{\boldsymbol{G}}_d \oslash \boldsymbol{G}_d.$$

Note that the operator $\oslash$ indicates elementwise division. A RDG entry with magnitude smaller than one indicates that the single loops support each other in rejecting the DV.

5.A.6. MSV

After simple algebra, equation (5.1) can be written as

$$\boldsymbol{c} = \boldsymbol{T}\,\boldsymbol{c}_{\mathrm{s}} - \boldsymbol{S}\,\boldsymbol{G}_d\,\boldsymbol{d} - \boldsymbol{T}\,\boldsymbol{n}^{\mathrm{c}}.$$

If control is active, it is $\boldsymbol{T} \approx \boldsymbol{I}$. Thus, $\boldsymbol{c}_{\mathrm{s}} = \boldsymbol{c} = \boldsymbol{G}\,\boldsymbol{u}$ which implies[4] $\|\boldsymbol{c}_{\mathrm{s}}\|_2 \leq \bar{\sigma}\left(\boldsymbol{G}\right)\,\|\boldsymbol{u}\|_2$. Analogously, one can write $\boldsymbol{u} = \boldsymbol{G}^{-1}\,\boldsymbol{c}_{\mathrm{s}}$ which implies $\|\boldsymbol{u}\|_2 \leq \bar{\sigma}\left(\boldsymbol{G}^{-1}\right)\,\|\boldsymbol{c}_{\mathrm{s}}\|_2 = \underline{\sigma}^{-1}\left(\boldsymbol{G}\right)\,\|\boldsymbol{c}_{\mathrm{s}}\|_2$. Combining these results yields

$$\frac{1}{\bar{\sigma}\left(\boldsymbol{G}\right)} \leq \frac{\|\boldsymbol{u}\|_2}{\|\boldsymbol{c}_{\mathrm{s}}\|_2} \leq \frac{1}{\underline{\sigma}\left(\boldsymbol{G}\right)}.$$

[4] Multiplicative analog of Weyl's inequalities (Horn and Johnson, 1990, Problem 18, pp. 423f).

Accordingly, a small minimum singular value (MSV) $\underline{\sigma}(\boldsymbol{G})$ indicates that the magnitude of control action may be undesirably large due to a setpoint change in a certain (bad) direction (Morari, 1983). Note that some authors refer to $\underline{\sigma}(\boldsymbol{G})$ as the Morari resiliency index (MRI). If $\boldsymbol{G}$ is appropriately scaled, the fact $\underline{\sigma}(\boldsymbol{G}) < 1$ indicates problems with MV constraints. Mostly unwanted is of course long term MV saturation, *i.e.*, $\underline{\sigma}(\boldsymbol{G}_0) < 1$, as setpoint tracking targets cannot be achieved at steady-state. It is interesting to note that the design of self-optimizing control structures favors large MSVs as can be seen by observation of (3.14).

5.B. PID controller tuning

Among other technologies, proportional-integral-derivative (PID) controllers are still most dominantly applied in industry (approx. 90 % share, Kano and Ogawa, 2009). However, there is a confusing spectrum of single-loop tuning rules for PID controllers available in literature. Åström and Hägglund (1995) give a comprehensive overview of PID technology and review the most established tuning rules such as the early one by Ziegler-Nichols. A more practical guide is presented by McMillan (2000). Single-loop PID tuning is to date a field of research as shown by recent works of Lee *et al.* (2000), Skogestad (2003), Kano and Ogawa (2009) and Shamsuzzoha *et al.* (2010/07/05-07), among others. A detailed comparison of advanced tuning rules is provided by Foley *et al.* (Aug 2005).

In this work, PID controller synthesis is based on setpoint tracking and disturbance rejection performance as discussed in the textbook of Skogestad and Postlethwaite (2005, pp. 448-451). According to them, good performance is achieved if

$$|1 + \boldsymbol{L}_{ii}| > \max_{p,q} \left\{ |\boldsymbol{\Gamma}_{ip}| , \left| \left(\tilde{\boldsymbol{G}}_d\right)_{iq} \right| \right\} \tag{5.8}$$

can be obtained. Satisfaction of (5.8) may conflict with both, stability requirements and the aversion to fast and strong MV affection. Considering SISO loops, it is well known that closed loop stability is satisfied if the condition (Bode's stability criterion, Skogestad and Postlethwaite, 2005, p. 27)

$$|\boldsymbol{L}_{ii}(j\,\omega_{180})| < 1 \tag{5.9}$$

is fulfilled, where $\angle \boldsymbol{L}_{ii}(j\,\omega_{180}) = -180\,°$. If ω_{180} is undetermined in the considered frequency band, then ω_{180} may be replaced by the bandwidth frequency ω_{b} as plant behavior is unpredictable beyond ω_{b} due to model uncertainties. Note that the margin between left and right hand side of (5.9) is referred to as the gain margin 1/GM, which is typically desired to be of magnitude 1/2 (Skogestad and Postlethwaite, 2005, p. 33).

The stated targets can be formulated in terms of an optimization problem for the tuning the ith PID controller loop. The proportional and the integral part, κ_i and τ_i, respectively, are obtained by solving

$$\{\kappa_i, \tau_i\} = \arg \min_{\kappa_i, \tau_i, c} c \tag{5.10a}$$

$$\text{s.t. } 0 \geq \max_{p,q} \left\{ |\boldsymbol{\Gamma}_{ip}(j\,\omega_k)|, \left| \left(\tilde{\boldsymbol{G}}_{\boldsymbol{d}}\right)_{iq}(j\,\omega_k) \right| \right\} - |\boldsymbol{L}_{ii}(j\,\omega_k)| - c\,\forall k \tag{5.10b}$$

$$0 \geq |\boldsymbol{L}_{ii}(j\,\omega_{180})| - 1 + \frac{1}{\text{GM}} \tag{5.10c}$$

for a certain frequency band of interest, *i.e.*, usually $\omega_k \in [0, \omega_{\text{b}}]\,\forall k$.

The solution of the controller synthesis problem (5.10a) through (5.10c) does not guarantee stability of the overall system, *i.e.*, $\text{Re}(\lambda_i(\boldsymbol{S})) < 0$, as loop interactions are not taken into account. According to (Skogestad and Postlethwaite, 2005, Theorem 10.3), overall stability is satisfied if the following three (necessary and sufficient) conditions hold simultaneously.

1. The plant $\boldsymbol{G}$ is stable.

2. Each individual loop $\boldsymbol{L}_{ii}$ is stable.

3. $\rho\left(\left(\boldsymbol{G} - \tilde{\boldsymbol{G}}\right)\tilde{\boldsymbol{G}}^{-1}\left(\boldsymbol{I} - \tilde{\boldsymbol{S}}\right)\right) < 1\,\forall\omega$, where ρ is the spectral radius.

The first is crucial for integrity and holds for plants considered in this work. The second is ensured by the controller synthesis constraint (5.10c). The third needs to be checked subsequently after the solutions to problem (5.10a) through (5.10c) have been obtained for all pairs.

Bibliography

K. J. Åström and T. Hägglund. *PID controllers: Theory, design, and tuning.* Instrument Society of America, Research Triangle Park, North Carolina, 1995. ISBN 1556175167.

E. Bristol. On a new measure of interaction for multivariable process control. *IEEE transactions on automatic control*, 11(1):133–134, 1966.

M. W. Foley, R. H. Julien, and B. R. Copeland. A Comparison of PID Controller Tuning Methods. *Canadian journal of chemical engineering*, 83(4):712–722, Aug 2005.

M. Hammer. Ph.D. thesis, Trondheim, Norway.

K. Havre. *Studies on controllability analysis and control structure design: Ph.D. thesis.* Ph.D. thesis, NTNU, Trondheim, Norway, Jan 1998.

R. A. Horn and C. R. Johnson. *Matrix analysis.* Cambridge University Press, Cambridge, UK, 1990. ISBN 0521386322.

M. Hovd and S. Skogestad. Simple frequency-dependent tools for control system analysis, structure selection and design. *Automatica*, 28(5):989–996, 1992.

M. Kano and M. Ogawa. The state of the art in advanced chemical process control in Japan. In S. Engell and Y. Arkun, editors, *ADCHEM 2009: Preprints of IFAC symposium on advanced control of chemical processes: July 12-15, 2009, Koç University, Istanbul, Turkey*, volume 1, pages 11–26. 2009.

H.-R. Lausch. *Ein systematischer Entwurf von Automatisierungskonzepten für komplex verschaltete Chemieanlagen: Ph.D. thesis*, volume 587 of *Fortschritt-Berichte VDI Reihe 8 Meß-, Steuerungs- und Regelungstechnik.* VDI-Verlag, Düsseldorf, Germany, 1999. ISBN 3183587033.

Y. Lee, J. Lee, and S. Park. PID controller tuning for integrating and unstable processes with time delay. *Chemical engineering science*, 55(17):3481–3493, 2000.

F. V. Lima, Z. Jia, M. Ierapetritou, and C. Georgakis. Similarities and differences between the concepts of operability and flexibility: The steady-state case. *American institute of chemical engineers journal*, 56(3):702–716, 2010.

J. A. Mandler and P. A. Brochu. Controllability analysis of the LNG process. AIChE annual meeting, Los Angeles, Californien, Nov 1997.

J. A. Mandler, P. A. Brochu, J. Fotopoulos, and L. Kalra. New control strategies for the LNG process. International conference & exhibition on Liquefied Natural Gas, Perth, Australia, May 1998.

T. J. McAvoy. *Interaction analysis: Principles and applications.* Instrument Society of America, Research Triangle Park, North Carolina, Sep 1983. ISBN 0876646313.

G. K. McMillan. *Good tuning: A pocket guide.* ISA, Research Triangle Park, North Carolina, 2000. ISBN 1556177267.

E. Melaaen. *Dynamic simulation of the liquefaction section in baseload LNG plants: Ph.D. thesis.* Ph.D. thesis, NTH, Trondheim, Norway, Oct 1994.

M. Morari. Design of resilient processing plants III: A general framework for the assessment of dynamical resilience. *Chemical engineering science*, 38(11):1881–1891, 1983.

M. Morari, E. Zafiriou, and B. R. Holt. *Chemical engineering science.*

A. Niederlinski. A heuristic approach to the design of interacting multivariable systems. *Automatica*, 7(6):691–701, 1971.

M. J. Okasinski and M. A. Schenk. Dynamics of baseload liquefied natural gas plants: Advanced modelling and control strategies. International conference & exhibition on Liquefied Natural Gas, Barcelona, Spain, 2007/04/ 24-27.

C. C. Pantelides, D. Gritsis, K. R. Morison, and W. H. Sargent. The mathematical modelling of transient systems using differential-algebraic equations. *Computers and chemical engineering*, 12(5):449–454, 1988.

B. C. Price and R. A. Mortko. PRICO: A simple, flexible proven approach to natural gas liquefaction. International LNG/LPG conference, Vienna, Austria, 1996/12/3-6.

M. Shamsuzzoha, S. Skogestad, and I. J. Halvorsen. On-line PI controller tuning using closed-loop setpoint responses for stable and integrating processes. International Symposium on Dynamics and Control of Process Systems, Leuven, Belgium, 2010/07/05-07.

A. Singh and M. Hovd. Mathematical modeling and control of a simple liquefied natural gas process. Scandinavian simulation society, Helsinki, Finland, 2006/09/28-29.

S. Skogestad. Simple analytic rules for model reduction and PID controller tuning. *Journal of process control*, 13(4):291–309, 2003.

S. Skogestad and M. Hovd. Use of frequency-dependent RGA for control structure selection. American Control Conference, San Diego, California, 1990/05/23-25.

S. Skogestad and M. Morari. Effect of disturbance directions on closed-loop performance. *Industrial and engineering chemistry research*, 26(10):2029–2035, 1987.

S. Skogestad and I. Postlethwaite. *Multivariable feedback control: Analysis and design.* Wiley, Chichester, UK, 2nd ed. edition, 2005. ISBN 0470011688.

G. Stanley, M. Marino-Galarraga, and T. J. McAvoy. Shortcut operability analysis 1: The relative disturbance gain. *Industrial and engineering chemistry research*, 24(24): 1181–1188, 1985.

E. A. Wolff. *Studies on control of integrated plants: Ph.D. thesis.* Ph.D. thesis, NTH, Trondheim, Norway, 1994.

A. Zaïm. *Dynamic optimization of an LNG plant: Case study GL2Z LNG plant in Arzew, Algeria: Ph.D. thesis*, volume 10 of *Schriftenreihe zur Aufbereitung und Veredelung*. Shaker, Aachen, Germany, Mar 2002. ISBN 3832206647.

G. Zapp. Dynamic simulation of air separation plants and the use of "Relative Gain Analysis" for design of control systems. *Linde reports on science and technology*, 54: 13–18, 1994.

Z.-X. Zhu and A. Jutan. RGA as a measure of integrity for decentralized control systems. *Transactions of the institution of chemical engineers*, 74(a):35–37, 1996.

www.ingramcontent.com/pod-product-compliance
Lightning Source LLC
Chambersburg PA
CBHW021047210326
41598CB00016B/1127
* 9 7 8 3 8 3 8 1 2 9 0 4 4 *